# Low Calorie and Special Dietary Foods

Editor

**Basant K. Dwivedi**
Vice President & Director
Research & Development and Quality Control
The Estee Corporation
Parsippany, New Jersey

CRC Press
Taylor & Francis Group
Boca Raton London New York

CRC Press is an imprint of the
Taylor & Francis Group, an **informa** business

First published 1978 by CRC Press
Taylor & Francis Group
6000 Broken Sound Parkway NW, Suite 300
Boca Raton, FL 33487-2742

Reissued 2018 by CRC Press



**Library of Congress Cataloging in Publication Data**

Main entry under title:
Low-calorie and special dietary foods.
Bibliography: p.
Includes index.
1. Food, Dietetic. 2. Low-calorie diet.
3. Nonnutritive sweeteners. I. Dwivedi, Basant K.
RM258.L68 613.2'8 77-23865
ISBN 0-8493-5249-5

A Library of Congress record exists under LC control number: 77023865

Publisher's Note
The publisher has gone to great lengths to ensure the quality of this reprint but points out that some imperfections in the original copies may be apparent.

Disclaimer
The publisher has made every effort to trace copyright holders and welcomes correspondence from those they have been unable to contact.

ISBN 13: 978-1-315-89509-3 (hbk)
ISBN 13: 978-1-351-07419-3 (ebk)

Visit the Taylor & Francis Web site at http://www.taylorandfrancis.com and the
CRC Press Web site at http://www.crcpress.com

# PREFACE

Eating is a basic biological drive. Its role is to sustain life and promote health. The fact that many people remain at or about the same weight for many years, even though their food intake and their energy expenditures vary considerably from time to time suggests that appetite is controlled, at least to some degree, by a physiological mechanism. It is also apparent that in people who tend to gain weight, this physiological control seems to be weaker as these people continuously overeat. It has been reported that in obesity-prone individuals, the physiological sensor(s) are seemingly much less sensitive to an energy intake that is too large than to one that is too small. They fail to direct the action necessary to reduce the intake of energy, and the percentage of fat in the body rises. They become exceedingly active, however, if there has been a loss of weight for any reason. Low calorie or reduced calorie, special dietary foods may help these people considerably in balancing their food intake and energy expenditure, and thus in controlling obesity.

Special dietary foods, including low calorie foods, are also needed for people with other abnormal physiological conditions, particularly for those suffering from inborn metabolic errors. The dietary needs of these people are discussed in the first chapter of *Low Calorie and Special Dietary Foods.*

Special dietary foods is an open and largely unexplored field. Studies with rats by Professor Jean Mauron of Nestlé and others have shown that biological aging may be the manifestation of naturally occurring enzymatic changes, suggesting that dietary composition plays an important role in aging. If these findings are found applicable to man, they stimulate additional research in this area to find a more biological approach to human welfare. It would also mean that food intake and composition will be adapted to different periods of life, tailoring the nutritional requirements to local environment, trying to control emotions, and preventing the deleterious effects of overwork and stress.

This book is based on the papers presented at the Symposium on Low Calorie and Special Dietary Foods at the annual meeting of the Institute of Food Technologists in Anaheim, California on June 8, 1976.

Thanks go to contributors for accepting the invitation to present papers in the Symposium and for adapting their presentations for publication in this book. I also want to take this opportunity to thank all of the individuals and organizations who have helped in exchange of ideas. Special thanks go to Charles H. Adler, The Estee Corporation, for his encouragement and for critically reviewing several of the papers.

B. K. Dwivedi
July 26, 1977

# THE EDITOR

**Dr. Basant K. Dwivedi** is Vice President and Director of Research & Development and Quality Control of The Estee Corporation, Parsippany, New Jersey.

Dr. Dwivedi received his B.S. from Panjab University, Chandigarh, India; his M.S. from the University of Saskatchewan, Saskatoon, Canada; and his Ph.D. from the University of Nebraska, Lincoln, Nebraska. In 1972, he joined the Department of Food Science at Cornell University, Ithaca, New York, as Research Associate. In 1974 he went to Booz-Allen & Hamilton, Inc. as Associate Research Director. In 1975, he joined The Estee Corporation in his present position.

Dr. Dwivedi has authored over 20 technical articles and authoritative reviews in the field of Food Chemistry and Technology. He has been the recipient of many honors, including the Panjab University Medal in 1967 and the Sigma Xi Award for Excellence in Research (Nebraska Chapter) in 1973. He is active in several scientific associations, including The American Chemical Society, The American Association for the Advancement of Science, and The Institute of Food Technologists (IFT). He was invited by the IFT to chair a symposium on Low Calorie and Special Dietary Foods at its annual meeting in June 1976. He is also active in a number of trade associations, such as Grocery Manufacturers of America and the National Confectionery Association. He has served on various technical committees, including service as chairman for the NAS/NRC Survey Committee on Flavoring Ingredients in Confections.

His current research interests include Human Nutrition, Physical and Chemical Interaction of Proteins, Fats & Carbohydrates and their Application in Food Research, Odor and Taste Perception, Flavor Chemistry, and Food Fermentations.

# CONTRIBUTORS

**Charles I. Beck, Ph.D.**
Manager of Advanced Technological Research
Kitchens of Sara Lee
Division of Consolidated Foods Corporation
Deerfield, Illinois

**Karl M. Beck, Ph.D.**
Manager,
Market Research and Development
Chemical and Agricultural Products Division
Abbott Laboratories
North Chicago, Illinois

**John J. Beereboom, Ph.D.**
Research Manager
Central Research
Pfitzer Inc.
Groton, Connecticut

**Basant K. Dwivedi, Ph.D.**
Vice President & Director
Research & Development and Quality Control
The Estee Corporation
Parsippany, New Jersey

**Hugh R. Linn, B.S.**
Consultant
Sommers, Gusick, Icardi
San Francisco, California

**Thomas F. Osberger, B.S.**
Technical Sales/Service Manager
Finn-Cal Sweetener Company
Deerfield, Illinois

**Leonard J. Vinson, Ph.D.**
Director of Biomedical Affairs
Lever Brothers Company
Edgewater, New Jersey

TABLE OF CONTENTS

Chapter 1

# FOOD PRODUCTS FOR SPECIAL DIETARY NEEDS

**B. K. Dwivedi**

## TABLE OF CONTENTS

# I. INTRODUCTION

At the recent National Nutritional Policy hearings, Dr. William E. Connor, Co-Chairman of the Select Committee's Panel on Nutrition and Health, testified:[1]

> The vast majority of Americans suffer from overabundance of food. The changed ecology of our land. . .has led to a whole new spectrum of diseases in which nutritional factors either play the prime etiological role or else are highly contributory to the development of the given disease state, that is coronary heart disease, obesity, and so on.

There is general agreement among nutritionists that most of the serious abnormal physiological conditions can by improved by proper diet modifications. The nature and magnitude of these benefits are estimated in Table 1. Besides disease prevention, it should be possible to enjoy good health, a longer active life span, and greater satisfaction in our day-to-day life from improved diets and nutrition.

In recent years, the utilization of formulated and fabricated foods has changed considerably. The traditional nutritional patterns no longer provide adequate nutrition to all segments of the population. Consequently, 1 in 3 men and 1 in 6 women in the U.S. are expected to die of heart disease; 25 million Americans suffer from high blood pressure, and about 5 million are afflicted by diabetes mellitus. Until recently, physicians largely ignored diet therapy as part of the treatment of these illnesses. More and more health professionals are now beginning to appreciate the fact that special diets and specially formulated foods could be effectively used in the management of abnormal physiological conditions.

The U.S. Food and Drug Regulations dealing with special dietary foods[3] define the foods for special dietary use as follows:

> (a) The term "special dietary use" as applied to food (including dietary supplements) used by man means a particular use for which an article purports or represented to be used, including but not limited to the following: (1) Supply a special dietary need that exists by reason of a physical, physiological, or other condition, including but not limited to the conditions of convalescence, pregnancy, lactation, infancy, allergic hypersensitivity to food, underweight, overweight, diabetes mellitus, or the need to control the intake of sodium . . .

It should always be remembered that dietary foods are foods and not medicines; as such, claims about their curative or preventive properties with regard to human health should not be made, as these claims often lead to confusion and misunderstanding in the consumers' minds. Regulatory agencies in most countries forbid any such claim about the dietary foods.

Two main categories of dietary foods are (1) Foods which meet the special nutritional needs of people suffering from abnormal physiological conditions such as deficiency states and metabolic disorders. Dietary foods in this category include foods with:

Low sodium content
Low gluten content
Low levels of certain amino acids
Low calorie content
High calorie content
Low fat content
Modified carbohydrates
Low protein content
High protein content

(2) Foods which are designed to meet normal physiological conditions such as those for infants and children, for pregnant and breast-feeding women, for the aged, and foods with supplementary nutrients required by intensive physical exertion or special environmental conditions.

In this chapter, only those foods designed for people suffering from abnormal physiological conditions (category 1) are discussed.

In hospitals, the formulation of a required special diet for the patient is frequently accomplished by deleting the undesirable foods from his meals. However, if the patient must follow the special diet for a long period of time or if he must change his eating habits for life, it is desirable that he not be denied certain foods or excluded from social meals. Under these conditions, specially formulated foods that fulfill the patient's nutritional needs and resemble normal food in appearance and flavor are highly desirable. The food industry has come a long way in fulfilling this need. However, the efforts manufacturers put into formulation and production of dietetic foods has been limited, primarily because of the limited market for these foods. Presently, increased consumer awareness of caloric content in the diet, has caused the food industry to take a second look at this market. It appears likely that within a decade,

TABLE 1

**Magnitude of Benefits from Nutrition Research**

| Health problem | Magnitude of loss | Potential savings from improved diet |
|---|---|---|
| | **Nutrition-related Health Problems** | |
| Heart and vasculatory | Over 1,000,000 deaths in 1967; over 5 million people with definite or suspect heart disease in 1960–1962 | 20% reduction |
| | $31.6 billion in 1962 | 20% reduction |
| Respiratory and infectious | 82,000 deaths/year | 20% fewer incidents |
| | 246 million incidents in 1967 | |
| | 141 million work days lost in 1965–1966 | 15–20% fewer days lost |
| | 166 million school days lost | 15–20% fewer days lost |
| | $5 million in medical and hospital costs | $1 million |
| | $1 billion in cold remedies and tissues | $20 million |
| Mental health | 2.5% of population of 5.2 million people are severely or totally disabled; 25 million people have manifest disability | 10% fewer disabilities |
| Infant mortality and reproduction | Infant deaths in 1967–79,000 | 50% fewer deaths |
| | Infant death rate 22.4/1,000 | 50% fewer deaths |
| | Fetal death rate 15.6/1,000 | 50% fewer deaths |
| | Maternal death rate 28.0/100,000 live births | 50% fewer deaths |
| | Child death rate (1–4 years) 96.1 per 100,000 in 1964 | Reduce rate to 10/100,000 |
| | 15 million with congenital birth defects | 3 million fewer children with birth defects |
| Early aging and lifespan | 49.1% of population, about 102 million people have one or more impairments | 10 million people without impairments |
| | People Surviving to Age 65 (%) | |
| | White males, 66 | 1% improvement per year to 90% surviving |
| | Negro males, 50 | |
| | White females, 81 | |
| | Negro females, 64 | |
| | Life expectancy (years) | |
| | White males, 67.8 | Bring Negro expectancy up to White |
| | Negro males, 61.1 | |
| | White females, 75.1 | |
| | Negro females, 68.2 | |
| Arthritis | 16 million people afflicted | 8 million people without afflictions |
| | 27 million work days lost | 13.5 million work days |
| | 500,000 people unemployed | 125,000 people employed |
| | Annual cost $3.6 billion | $900 million/year |
| Dental health | 44 million with gingivitis; 23 million with advanced periodontal disease; $6.5 billion public and private expenditures on dentists' services in 1967; 22 million endentulous persons (1 in 8) in 1957; 1/2 of all people over 55 have no teeth | 50% reduction in incidence, severity, and expenditures |
| Diabetes and carbohydrate disorders | 3.9 million overt diabetic; 35,000 deaths in 1967; 79% of people over 55 with impaired glucose tolerance | 50% of cases avoided or improved |

TABLE 1 (continued)

**Magnitude of Benefits from Nutrition Research**

| Health problem | Magnitude of loss | Potential savings from improved diet |
|---|---|---|
| | **Nutrition-related Health Problems (continued)** | |
| Osteoporosis | 4 million severe cases; 25% of women over 40 | 75% reduction |
| Obesity | 3 million adolescents; 30–40% of adults; 60–70% over 40 years | 80% reduction in incidence |
| Anemia and other nutrient deficiencies | See entries for Improved work efficiency, Improved growth and development, and Improved learning ability | |
| Alcoholism | 5 million alcoholics, 1/2 are addicted | 33% |
| | About 24,500 deaths in 1967 caused by alcohol | 33% |
| | Annual loss over $2 billion from absenteeism, lowered production, and accidents | 33% |
| Eyesight | 48.1% or 86 million over 3 years wore corrective lenses in 1966; 81,000 become blind every year; $103 million in welfare | 20% fewer people blind or with corrective lenses |
| Cosmetic | 10% of women ages 9 or more with vitamin intakes below recommended daily allowances | |
| Allergies | 32 million people (9%) are allergic | 20% people relieved |
| | 16 million with hayfever asthma | — |
| | 7–15 million people (3–6%) allergic to milk | 90% people relieved |
| | Over 693 thousands persons (1 in 3,000) allergic to gluten | 90% people relieved |
| Digestive | 8,495 workdays lost; | 25% fewer acute conditions |
| | 5,013 school days lost; | 25% fewer acute conditions |
| | About 20 million incidents of acute condition annually | 25% fewer acute conditions |
| | $4.2 billion annual cost; 14 million persons with duodenal ulcers; $5 million annual cost; 4,000 new cases each day | Over $1 billion in costs |
| Kidney and urinary | 55,000 deaths from renal failure; 200,000 with kidney stones | 20% reduction in deaths and acute conditions |
| Muscular disorders | 200,000 cases | 10% reduction in cases |
| Cancer | 600,000 persons developed cancer in 1968; 320,000 persons died of cancer in 1968 | 20% reduction in incidence and deaths |
| | **Individual Satisfactions Increased** | |
| Improved work efficiency | | 5% increase in on-the-job productivity |
| Improved growth and development | 113,000 deaths from accident, 324.5 million workdays lost, 51.8 million people needing medical attention and/or restricted activity | 25% fewer deaths and workdays lost |

TABLE 1 (continued)

**Magnitude of Benefits from Nutrition Research**

| Health problem | Magnitude of loss | Potential savings from improved diet |
|---|---|---|
| | **Nutrition-related Health Problems (continued)** | |
| Improved learning ability | Over 6.5 million mentally retarded persons with I.Q. below 70; 12% of school age children need special education | Raise I.Q. by 10 points for persons with I.Q. 70–80 |

From Weir, C. E., Nutrition and Health, With An Evaluation of Nutritional Surveillance in the United States, U.S. Government Printing Office, Washington, D.C., 1975.

dietetic foods, particularly low calorie foods, will be one major category of processed foods.

## II. LOW SODIUM FOODS

Sodium is involved in maintenance of body fluid pH, osmolality, and body fluid volume. These functions are accomplished by the unequal distribution of sodium and potassium ions in the plasma membrane. A higher concentration of potassium is found inside the membrane compared to outside, the reverse is true for sodium. The system is commonly known as $K^+/Na^+$ pump. The "pump" is universally found in the plasma membrane of cells where it is localized. Under the following conditions, the sodium concentration in the plasma membranes increases, resulting in fluid retention or edema in the body:[4]

- Nephritis
- Nephrosis
- Cardiac disease
- Cirrhosis of the liver
- Toxemia of pregnancy
- Hypertension
- Adrenocarticotrophin therapy

In these conditions, the most successful treatment for prevention of further edema is restriction of sodium content in the diet.

Daily sodium intake, mostly in the form of sodium chloride, depends on eating habits and varies considerably from person to person. Although only about 0.5 g of sodium per day is required for proper biological functions, the average sodium intake comes to about 3 to 6 g/day. The salt or sodium content in the diet should be restricted as required by the severity of disease and edema, using the following guidelines.

**Mild restriction** – In the case of mild heart damage, the intake should be restricted to 2 to 3 g/day. Deletion of highly salted foods from the diet is sufficient to achieve this level. Salting cooked foods lightly may be permitted under this condition.

**Moderate restriction** – If the patient has edema, the sodium intake should be restricted to about 1 g/day. To achieve this level, it is necessary to delete salt (sodium chloride) completely from the diet and permit only the foods which are low or moderate in sodium.

**Severe restriction** – In the case of congestive heart failure, severe edema, and hypertension, the sodium intake should be restricted to about 0.5 g/day. This level is difficult to achieve, and it is necessary that only foods low in sodium and the specially formulated low sodium foods be permitted.

Commercially, low sodium foods are prepared by using ingredients inherently low in sodium Removal of sodium from ingredients during processing, although feasible, is not practiced, probably because of excessive costs. A recently perfected ultrafiltration process may lend itself to removal of sodium from solutions and colloidal suspensions at a reasonable cost. The prime capability of ultrafiltration lies in the membrane's ability to pass or reject molecular species on the basis of differences in shape, size, and chemical structure. Since sodium salts comparatively have a much lower molecular weight than carbohydrates and proteins, they are easily removed by ultrafiltration. In preparing low sodium foods, the following substances may be used to replace sodium chloride:

I. Calcium, magnesium,* and potassium salts of these acids
   - Acetic
   - Adipic
   - Carbonic
   - Citric
   - Hydrochloric
   - Glutamic
   - Tartaric
   - Succinic

II. Potassium sulfate

III. Choline salts of these acids
   - Acetic
   - Carbonic
   - Citric
   - Hydrochloric
   - Lactic
   - Tartaric

IV. Adipic and glutamic acids

## III. LOW GLUTEN FOODS

Low gluten foods are recommended in coeliac disease. This disorder is characterized by poor absorption of fat and other nutrients, abdominal distension, and loss of weight. The malabsorption of nutrients often leads to anemia from iron deficiency, tetany from calcium deficiency, and multiple vitamin deficiency.

In coeliac disease, intestinal walls of the susceptible persons are damaged through the ingestion of wheat and rye proteins. Since gluten is the major functional protein in these cereals, the malabsorption syndrome is also referred to as gluten-induced enteropathy.

The texture of baked products such as bread, cakes, and biscuits depends on gluten to a large extent. Gluten also plays an important role in the texture and mouthfeel of macaroni, noodles, spaghetti, and other cereal foods. The exclusion of wheat flour and gluten from the baked foods formulations causes severe limitations, as gluten has unique functional properties which are difficult to replace. Soy protein and casein are often used as replacements for gluten in baked foods formulations. However, the dough prepared with gluten replacements lacks elastic character and results in poor texture after baking. Egg albumin appears to be a good protein for imparting the textural strength to baked foods in the absence of gluten. Xanthane gum has also been suggested as a partial replacement of gluten.

## IV. FOODS LOW IN PHENYLALANINE

A number of metabolic defects are related to abnormal phenylalanine metabolism. These disorders, as illustrated in Figure 1, include the following.

**Phenyl ketonuria** – Under this condition, the conversion of phenylalanine to tyrosine is blocked. Persons suffering from this disorder lack the catalyzing enzyme, mixed function oxygenase.

**Tyrosinosis** – The conversion of *p*-hydroxyphenyl pyruvate to homogentisic acid through an intermediate, quinol, is blocked because of the lack of enzyme oxygenase.

**Alkaptonuria** – The conversion of homogentisic acid to acetoacetic acid and fumaric acid is blocked. Persons suffering from this condition cannot fully metabolize phenylalanine and tyrosine, and homogentisic acid is excreted in the urine.

Of the above three, phenyl ketonuria is the most serious inborn error of metabolism. The abnormal products of phenylalanine metabolism – phenylacetic acid, phenyllactic acid, and phenylacetylglutamine – damage the brain beyond repair. If the disease is detected within the first few weeks of life before any brain damage occurs, strict control of phenylalanine in the diet and the blood allows the child to develop normally.

Treatment of this disorder involves the restriction of phenylalanine in the diet. A small amount of phenylalanine is permitted in the diet for protein synthesis. Management of the disease consists of supplying only the bare minimum amount so that there is no surplus phenylalanine to accumulate and be converted to harmful metabolites.[4]

Low phenylalanine foods are prepared by hydrolyzing a protein such as casein and passing it through an activated charcoal column which traps phenylalanine. In this process, tryptophan, tyrosine, and cystine are also lost. These amino acids are replaced with commercially available amino

*Magnesium salts of the acids could be used only in limited quantity; usually the magnesium content should be less than 20.0% by weight of the total cations present in the mixture.

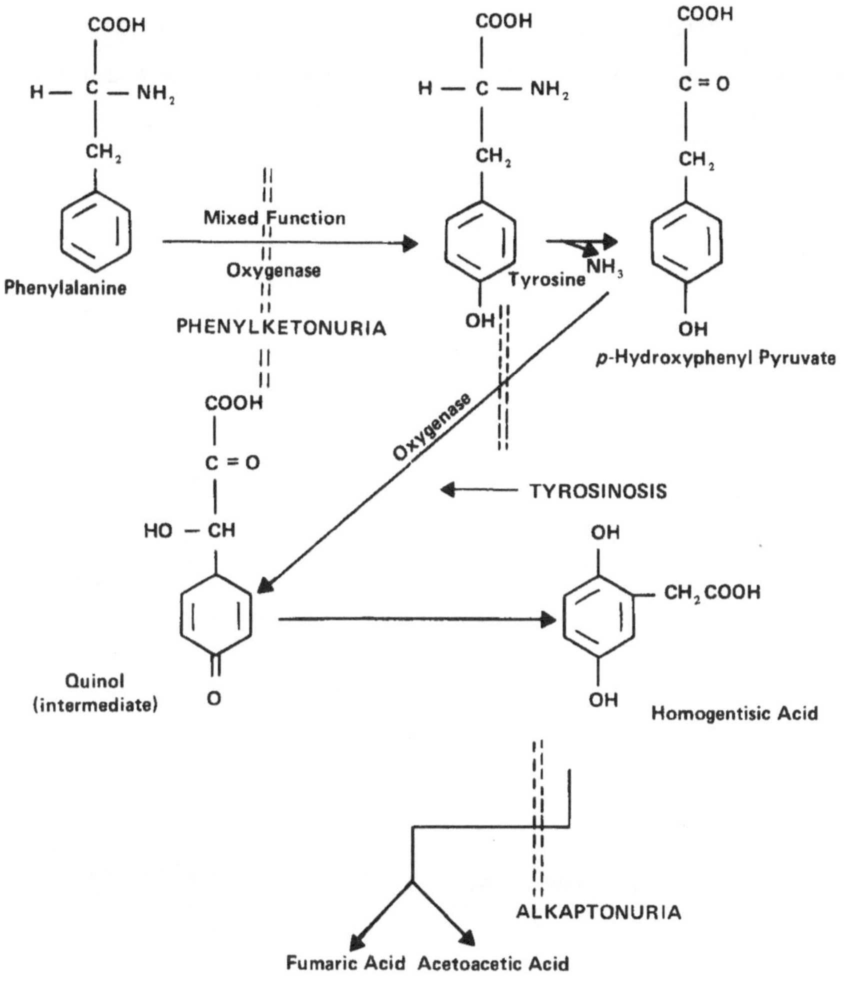


FIGURE 1. Phenylalanine-related inborn errors of metabolism.

acids. This phenylalanine-free hydrolysate is used as a base for preparing low phenylalanine foods. Formulation of palatable foods from the hydrolysate is a difficult task because the hydrolysis releases free amino acids, some having an unpleasant taste as shown in Table 2. Approaches that should be investigated for making the products with protein hydrolysate more palatable include (1) coating or encapsulation of hydrolyzed protein solids, (2) adsorption of amino acids on a suitable adsorbant, and (3) weak interaction with other food grade materials.

## V. FOODS WITH LOW CALORIE CONTENT

Overweight or obesity is an abnormal physiological condition. It places a considerable strain on the biological system of the organism and often leads to diabetes mellitus, cardiovascular illness, and other diseases. Weight reduction is also important in the treatment of renal, gout, and gall bladder diseases.

Low-caloric foods designed for weight reduction may be formulated by using noncaloric sweeteners as replacement for sugars in jams, canned fruits, and confectionery products (see Chapter 5, Application Potential for Aspartame in Low Calorie and Dietetic Foods) and by using low-caloric bulking agents such as cellulose derivatives, pectins, vegetable gums, agar, etc. This subject is covered in Chapter 3, Low Calorie Bulking Agents. Lowering of fat in foods also causes reduction in caloric density. This aspect is discussed below in Section VI.

TABLE 2

**Organoleptic Characteristics of Amino Acids**

| Amino acid | Relative taste intensities | | |
|---|---|---|---|
| | Sweet | Sour | Bitter |
| Hydroxyproline | *** | | ** |
| Lysine-HCl | ** | | ** |
| Alanine | *** | | |
| Glycine | *** | | |
| Serine | *** | * | |
| Glutamine | * | | |
| Threonine | *** | * | * |
| Proline | *** | | *** |
| Aspartic acid | | *** | |
| Glutamic acid | | *** | |
| Histidine-HCl | | *** | |
| Asparagine | | ** | * |
| Histidine | | | ** |
| Arginine-HCl | | | *** |
| Methionine | | | *** |
| Valine | | | *** |
| Arginine | | | *** |
| Isoleucine | | | *** |
| Phenylalanine | | | *** |
| Tryptophan | | | *** |
| Leucine | | | *** |

*Note:* Asterisks denote relative taste intensities: * least taste intensity; ** medium taste intensity; *** most taste intensity.

Adapted from Kirimura, J., Shimizu, A., Kimizuka, A., Ninomiya, T., and Katsuya, J., *J. Agric. Food Chem.*, 17, 689, 1969.

## VI. FOODS WITH HIGH CALORIE CONTENT

High calorie foods are recommended during prolonged illness which leads to loss of body weight and undernutrition. Hyperthyroidism and fever also cause weight loss because of increased metabolic rate. The metabolic rate increases by 7% for each degree (Farenheit) rise in body temperature.[4]

High calorie foods may be prepared by increasing the fat and carbohydrate content at the expense of poorly digestible polysaccharides and moisture. It is important to remember that during illness, the body loses its reserve nutrients, so high calorie foods should be nutritionally balanced as much as possible.

TABLE 3

**Comparative Nutritional Analysis Egg vs. a Commercial Egg Substitute**

| | Egg | Egg substitute |
|---|---|---|
| Serving size, g | 47 | 47 |
| Protein, g | 5.62 | 4.65 |
| Lipids, g | 5.26 | 1.60 |
| polyunsaturated | 0.66 | 0.77 |
| monounsaturated | 2.91 | 0.55 |
| saturated | 1.69 | 0.28 |
| Cholesterol, mg | 259 | Trace |
| Carbohydrate, g | <1 | <1 |
| Calories | 78 | 35 |

Adapted from Anon., *Food Prod. Dev.*, p. 12, June 1975.

## VII. FOODS WITH LOW FAT CONTENT

Low fat foods are recommended for gall bladder disease to rest the gall bladder and for coeliac disease. Since fat is the most concentrated form of energy, low fat foods are desired in obesity (see Chapter 2, Dietary Lipids in Health and Disease).

A number of products with low fat content are commercially available. Typical low calorie, low fat products include salad dressings, egg substitutes and margarines. In salad dressings which contain as low as 10% oil, the desirable unctuous quality, attributable to fat in a regular dressing containing 30 to 50% fat, is compensated for by the incorporation of sodium carboxymethylcellulose (CMC). CMC contributes thickening, stabilizing, and emulsifying properties to the dressing. Egg substitutes in which the yolk fraction is replaced with specialty fats, caseinate, vitamins, and minerals to cut calories and cholesterol are becoming very popular. With the exception of quantity and type of fat, substitutes contain a higher concentration of polyunsaturated fatty acids. Compared to egg, the egg substitutes provide nearly identical nutrients (Table 3).

It is heartening to note that with the availability of products such as 99% fat-free milk, cholesterol-free and reduced calorie egg products, and polyunsaturated low calorie margarine, dieting is becoming an easier task for the average consumer. This, however, is still a wide-open field with many product opportunities such as genetically reduced-cholesterol eggs, leaner beef, pork, and lamb, formulated and prepared foods, snack products, and pastry products using polyunsaturated oils.[6]

## VII. FOODS WITH MODIFIED CARBOHYDRATES

Approximately 50% of our diet consists of carbohydrates and provides about 48 to 55% of total energy. Of the total carbohydrates consumed, it is estimated that up to 60% are derived from potatoes, corn, rice, or bread. Sucrose in the form of table sugar and in desserts accounts for another 30%. Lactose from milk, ice cream, and other dairy products accounts for the remaining 10%.[7]

Under certain abnormal physiological conditions, utilization of some carbohydrates is affected. Disaccharides, for example, are incompletely absorbed and utilized if the appropriate intestinal enzyme such as sucrase, maltase, or lactase is lacking or deficient. In diabetics, glucose metabolism is impaired due to lack of insulin, thus increased blood glucose level results. The enzyme deficiency can occur in severe protein malnutrition (kwashiorkor) or in diseases of the intestinal tract where the patient cannot tolerate wheat protein, gliadin.[7]

### A. Lactose Intolerance

Among the three disaccharide hydrolases mentioned above, lactose deficiency is most serious (Table 4). Some 70% of the world's adult population is affected by it. This genetic disorder is characterized by the fact that affected persons are able to drink milk as infants, but gradually lose the capacity to utilize lactose because of low levels of intestinal lactase activity. The enzyme lactase is formed within the brush border of microvillous area of the jejunal mucosa and is required for the hydrolysis of lactose into glucose and galactose (Figure 2). In lactose-intolerant subjects, ingested lactose is not hydrolyzed because of the low lactase activity. This results in abdominal distention, cramps, and increased peristalsis caused by the fluid that enters the intestinal lumen to dilute the hypertonic load of lactose.[9]

TABLE 4

**Prevalence of Lactase Deficiency in Healthy Population**

| Group | No. tested | Location | Lactose deficient (%)[a] |
|---|---|---|---|
| | **By Intestinal Biopsy Assay** | | |
| White | 380 | U.S. | 18 |
| | 700 | Denmark | 3 |
| | 14 | Puerto Rico | 14 |
| American Negro | 44 | U.S. | 73 |
| African Bantu | 40 | Uganda | 50 |
| Puerto Rican | 28 | Puerto Rico | 21 |
| Oriental | 11 | U.S. | 100 |
| | **By Oral Lactose Tolerance** | | |
| White | 23 | Australia | 0 |
| | 17 | Greek Cypriots | 88 |
| Aborigine | 44 | Australia | 85 |
| Chami Indian | 24 | Columbia | 100 |
| Chinese | 15 | Australia | 87 |
| | 10 | U.S. | 95 |
| Eskimo | 32 | Greenland | 88 |
| Indian | 22 | North India | 55 |
| Filipino | 10 | Baltimore, Md. | 95 |
| Thai | 140 | Thailand | 97 |

[a]Lactase activity below 1.0 U/g tissue or 15 U/g protein.

Adapted from Gray, G. M., *Annu. Rev. Med.*, 222, 391, 1971.

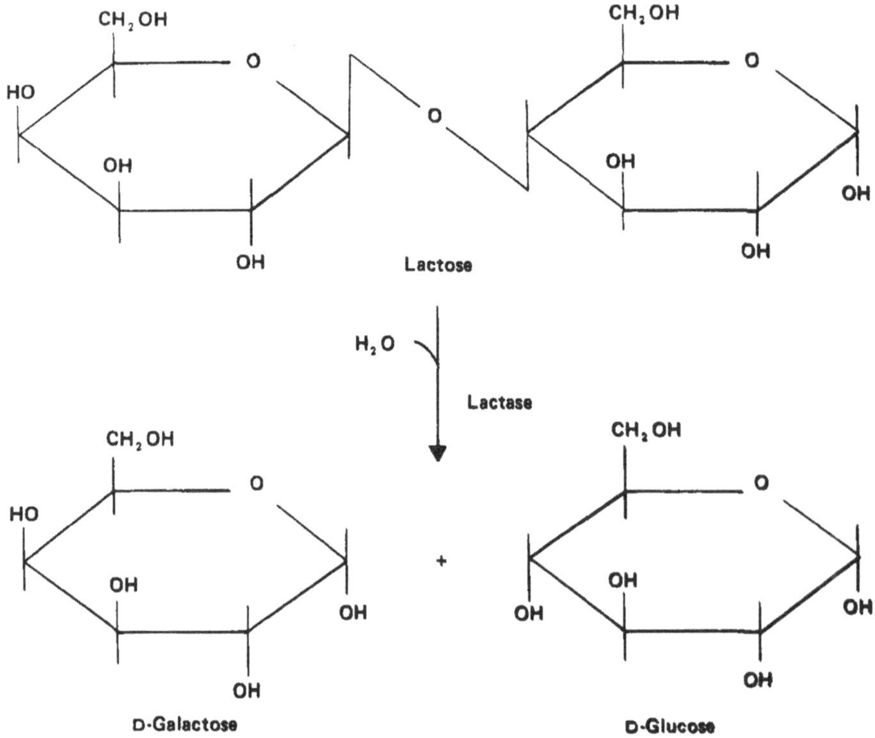


FIGURE 2. Conversion of lactose to D-galactose and D-glucose with lactase.

Paige et al.[9] studied the lactose utilization in lactose-intolerant subjects using whole milk and lactose-hydrolyzed milk. In the intolerant group, the difference between untreated and 90% hydrolyzed milk was found significant at the 0.001 level and between 50 to 90% hydrolyzed milk at the 0.02 level. This study shows that for lactose-intolerant subjects, the appropriate level of lactose hydrolysis should approach 90% as significant differences in the blood sugar rise were noted between 50 and 90% hydrolysis.

Commercial lactose preparations are becoming available for the hydrolysis of milk lactose.[10] One such preparation, when added to milk, converts over 70% of lactose to its components, glucose and galactose, in 24 hr at refrigeration temperature (~ 4°C). Higher conversion could be achieved by adding a higher amount of the lactose preparation.

### B. Galactosemia

Galactosemia is an inborn error of metabolism caused by uridyltransferase deficiency which normally catalyzes the second step in the conversion of galactose to glucose (Figure 3). This results in cataracts, liver and kidney malfunction, and disturbed mental development caused by organic brain damage. Under this condition, galactose-1-phosphate accumulates when galactose is ingested. The exclusion of galactose sources from the diet of galactosemics is usually recommended for the management of the disease. Under controlled conditions, levels of galactose-1-phosphate lower than 4 mg/100 ml are maintained in most patients.[11] However, in some galactosemics, uncontrolled biosynthesis of galactose-1-phosphate has been observed, even when their intake of galactose source is carefully regulated. Since milk and dairy products are the only major sources of galactose, deletion of these products from the diet usually brings the disease under control.

### C. Dental Caries

A British panel on cariogenic foods, Committee on Medical Aspects of Food Policy[12] defined dental caries as follows:

Dental caries is a disease which starts at the surface of the tooth, it penetrates the enamel and dentine, and if

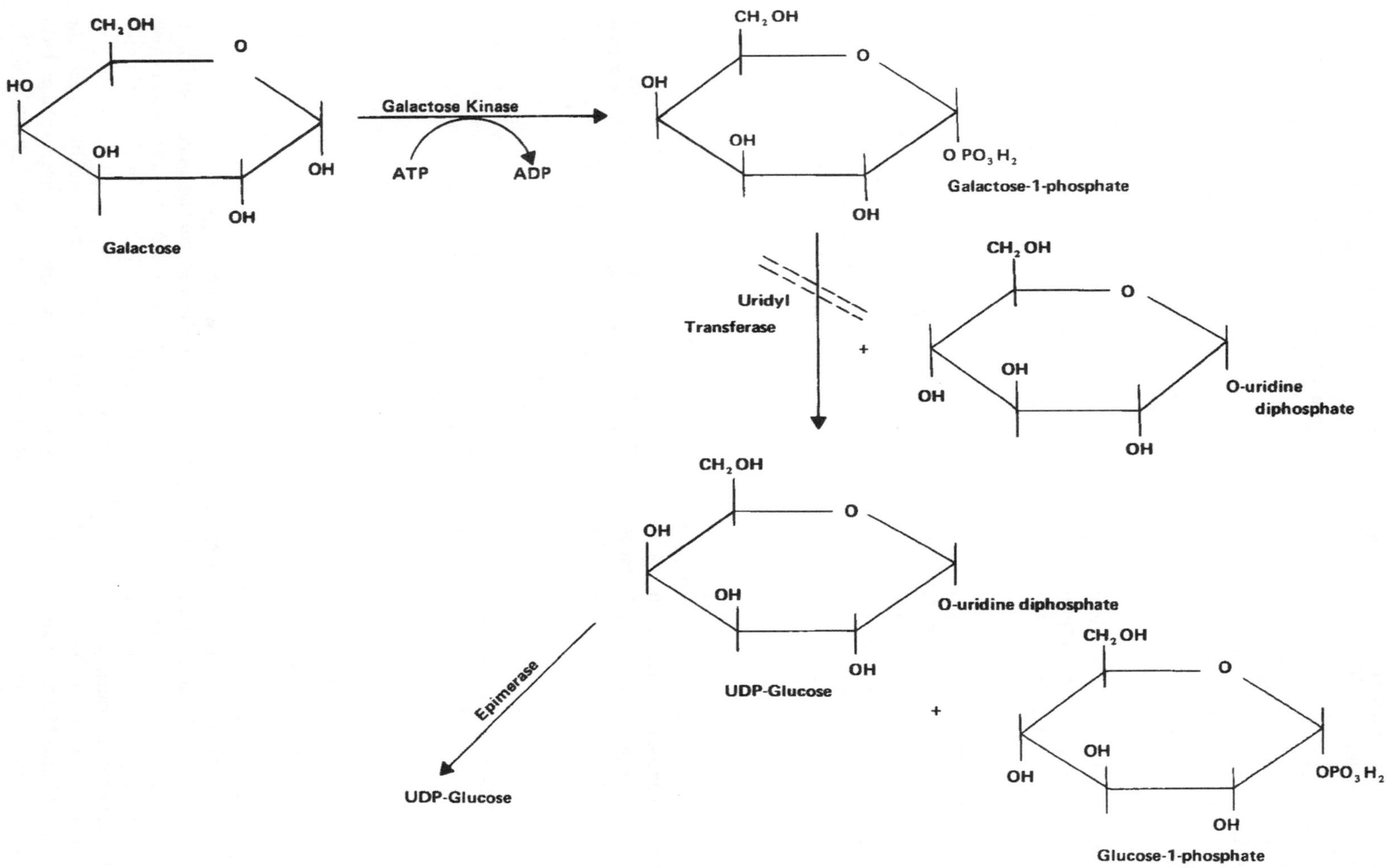


FIGURE 3. Initial steps in galactose metabolism.

unchecked will eventually reach the dental pulp destroying the tooth. The onset of initial lesion is due to demineralization by the acid secretions of microorganisms on the enamel surface of the tooth in a structure known as the dental plaque. Incontrovertible evidence exists that the micro-organisms of the plaque can convert fermentable carbohydrates into organic acids and it is these acids which lower the pH locally to a point where dissolution of the highly mineralized enamel cover occurs, causing the initial break on the surface of the enamel.

Five factors essential in the development of human caries include (1) a caries-susceptible individual, (2) the presence of acid producing and acid tolerating microorganisms in the oral cavity, (3) the presence of acid resistant microorganisms capable of producing an optimum oral bacterial enzyme system for the prompt degradation of simple sugars to acid, (4) the availability of orally-fermentable carbohydrates, and (5) bacterial plaque adherent to the surface of the tooth enamel. Although many oral microorganisms may be considered cariogenic, experimental observations indicate that some bacterial species are more cariogenic than others on certain tooth surfaces.[14]

Sucrose is probably the most cariogenic common sugar. The high energy of hydrolysis of the covalent bond between the monosaccharide of sucrose apparently helps in caries development. The energy released from the sucrose hydrolysis is used by oral microflora in the synthesis of hydrophilic polysaccharides which may form a bulk of the extracellular compartment of dental plaque. The polysaccharides produced, dextrans and levans, partly promote the growth and manifestation of cariogenic and other bacteria resulting in diffusion barriers. Dextrans and levans act as inflammatory agents, and provide starting material for the formation of simple cariogenic monosaccharides, glucose and fructose, and acid.[14]

Stephan[15] investigated the cariogenicity and anticariogenicity of various foods and beverages in depth. The most common factors associated with the caries development were a history of frequent or excessive eating of sugar-based products during and between meals, extensive dental plaque formation, and lower pH levels in plaques and food particles retained in caries areas than in mucous membranes or saliva. Carefully controlled experiments with Osborne-Mendel strain rats were also conducted by Stephan[15]. The basic cariogenic diets consisted of fine particles low in fat and high in sugar, and the noncariogenic diets were essentially the same except that the sucrose was deleted. Fifty-three representative foods and refreshments were tested for cariogenicity. Caries scores for the cariogenic and noncariogenic diets are listed in Table 5. Of the foods tested, cola, apples, orange and lemon slices, grapes, dried apricots, and orange drink produced dental erosion. These foods were quite acidic, with a pH lower than 4.5. Generally, the data showed that habitual and excessive consumption of sugar-based products may lead to the development of caries in susceptible individuals.

Mackinen[14] published an extensive review on the role of sucrose and other sugars in development of dental caries. Based on animal and human studies, Mackinen concluded that the cariogenicity of some common sugars decreases roughly in the following order: sucrose> glucose> maltose, lactose> fructose> sorbitol> xylitol.

Based on these studies, the American Dental Association recently issued a policy statement:[16]

> It is generally accepted that the frequent use of sucrose is the major cause of dental decay. The presence of sucrose in confections, such as chewing gum and candy mints, pose a particular problem since these products are used by persons of cavity prone ages .... There are, however, viable alternatives to sucrose-containing confections. These products contain hexitol, such as sorbitol and mannitol, or a pentitol, i.e. xylitol, with or without artificial sweeteners. They contain no sucrose. Products of this nature, under existing opinions of the American Dental Association are offered to consumers as non-cariogenic, i.e. they do not promote tooth decay. The non-cariogenic claim is justifiable when used in connection with such products because, unlike similar products which contain sucrose, they do not contribute to the development of dental caries ...

**D. Diabetes Mellitus**

Diabetes mellitus is a metabolic disorder, attributable to a deficiency of insulin or a complete lack of it. It results in insufficient peripheral glucose assimilation and an increase in glucose synthesis. The increased gluconeogenesis in diabetes may be accounted for by the lack of inhibition of the key gluconeogenetic enzymes. This situation is quite similar to that which exists in the fasting state. In addition, lack of insulin results in the lack of key glycolytic enzyme stimulation.[17] As in the state of hunger, the diabetic organism tries to circumvent the problem of inadequate peripheral glucose assimilation by using protein, glycerol, lactate, and pyruvate for the synthesis of glucose. The basic difference

TABLE 5

Comparison of Caries Scores (Mean Values) Produced by Different Foods

| | Experiments with noncariogenic supplement (Basic diet 581) | Experiments with cariogenic supplement (Basic diet 580) |
|---|---|---|
| Dog biscuits | 0.0 | 0.0 |
| Popcorn | 0.0 | 1.8 |
| Sorbitol | 0.0 | 1.0 |
| Peanuts | 0.0 | 2.3 |
| Whole milk[a] | 0.0 | 4.2 |
| Lemons[b] | 0.0 | 4.3 |
| Corn chips | 0.0 | 7.0 |
| Cabbage | 0.0 | 7.0 |
| Lettuce | 0.0 | 9.7 |
| Control (basic diet only 2-hr supplement | 0.0 | 11.3 |
| Dried apricots[b] | 0.0 | 11.3 |
| Oranges[b] | 0.0 | 16.8 |
| Soda crackers | 0.3 | 11.6 |
| Spinach | 0.6 | 4.6 |
| Cracked wheat bread | 1.3 | 13.7 |
| Potato chips | 1.6 | 6.7 |
| Whole wheat bread | 2.0 | 17.7 |
| Carrots[a] | 2.1 | 21.1 |
| Cornstarch[a] | 3.3 | 10.2 |
| White bread and butter | 4.9 | 22.1 |
| White bread and peanut butter | 5.2 | 18.0 |
| Graham crackers | 8.7 | 22.7 |
| Raisin bread | 9.0 | 17.2 |
| Melba toast | 9.0 | 19.6 |
| White bread | 9.2 | 15.1 |
| White bread and raspberry jam | 10.2 | 33.0 |
| Figs | 10.3 | 29.0 |
| Rye bread | 12.7 | 29.0 |
| Chewing gum | 14.0 | 27.2 |
| Caramels | 16.0 | 50.4 |
| Chocolate graham crackers | 18.0 | 39.1 |
| Baby cookies | 18.5 | 22.0 |
| Honey graham crackers | 19.2 | 21.6 |
| Apples[b] | 19.4 | 81.7 |
| Vanilla wafers | 19.7 | 30.2 |
| Bananas | 21.0 | 73.7 |
| Chocolate sandwich cookies | 23.8 | 92.2 |
| Grapes[b] | 24.1 | 42.4 |
| Candy mints | 24.7 | 42.12 |
| Teething biscuits | 29.0 | 29.2 |
| Cola[b] | 29.6 | 40.2 |
| Marshmallows | 30.1 | 42.2 |
| Dextrose | 30.6 | 43.5 |
| Raisins | 30.9 | 61.6 |

[a]Produced dental erosion.

[b]Produced a higher than average calculus score >1.5.

TABLE 5 (continued)

Comparison of Caries Scores (Mean Values) Produced by Different Foods

| | Experiments with noncariogenic supplement (Basic diet 581) | Experiments with cariogenic supplement (Basic diet 580) |
|---|---|---|
| 10% Sucrose water | 32.3 | 20.6 |
| Dates | 32.7 | 38.5 |
| Milk chocolate | 34.1 | 41.2 |
| White bread and jelly | 36.7 | 36.2 |
| Confectioners' sugar | 38.4 | 60.2 |
| 10% Dextrose water | 41.3 | 45.5 |
| Orange drink[b] | 43.5 | 26.1 |
| Sucrose | 62.1 | 51.9 |

Adapted from Stephan, R. M., *J. Dental Res.*, 45, 1551, 1966. With permission.

between gluconeogenesis in the state of hunger and gluconeogenesis in diabetes is that this process occurs under the hyperglycemic condition, in the case of insulin deficiency whereas in the state of hunger, gluconeogenesis occurs at a normal or slightly below normal blood glucose level.

Because of insulin deficiency, marked increase in lypolysis is observed in diabetics. The increased lypolysis, in turn, leads to an increase of free-fatty acids in serum. In addition, the triglyceride synthesis in fatty tissue is impaired by the lack of glycerophosphate. The excess free-fatty acids and reduced assimilation of acetyl-CoA results in an increase in ketone bodies, i.e., β-hydroxybutyric acid, acetic acid, and acetone. These ketone bodies slowly accumulate in the organism and overtax the reserves of alkali, eventually leading to acidosis, precoma, or coma diabeticum.[17] As a result of increased gluconeogenesis from protein in diabetes mellitus, the protein metabolism is also impaired.

The excess blood sugar in diabetes results in glucosuria and polyurea. The weakness and temporary weight loss observed in diabetic patients is apparently caused by the disturbed protein synthesis and the loss of body protein, loss of glucose, electrolyte and water in urine, and the catabolism of fatty tissues and resultant ketosis.[17]

Recent studies in dogs, rats, monkeys, hamsters, mice, and other animals have shown that reduction of hyperglycemia by insulin therapy, by transplanting insulin-producing tissues, or by some other means, prevents or minimizes the formation of eye, kidney, and nerve lesions associated with diabetes.[18] Persistent hyperglycemia has been shown to be associated with accumulation of sorbitol in nerve, eye, and vascular tissue and with alterations in vascular basement membrane.[19] The American Diabetes Association recently summarized current clinical and experimental findings on diabetes management and stressed that the goal of appropriate therapy should be to achieve optimal blood glucose regulation in the treatment of diabetes, particularly in young and middle-aged individuals who have a greater risk of developing microvascular complications.[20] The American Diabetes Association also concluded that current means of therapy are only partly effective; therefore, a high priority must be assigned to development of more physiological insulin delivery systems or to correction of the deficient insulin-producing mechanism itself. However, studies carried out by Shabo[21] suggest that exogenous insulin may contribute to the development of diabetic proliferative ocular disease. If these findings are substantiated by further studies, it will be necessary to de-emphasize exogenous insulin treatment and find alternate means to control blood glucose level.

Diet is the cornerstone of therapy for diabetic patients. Diabetic diet is calculated on the basis of 30 cal/kg of ideal body weight, using 40% carbohydrates, 40% fat, and 20% protein.[22] Diabetic patients are also advised to exclude the quickly absorbable carbohydrates, dextrose, invert sugar, and disaccharides from their diet. The exclusion of mono disaccharides and disaccharides from the diet of diabetics deprives them of one of the most satisfying diet components – desserts and sweets. It is frequently assumed that a great majority of diabetics recognize their problem and promptly discontinue sweets at their physician's direction. However, a poll conducted among 500 diabetics

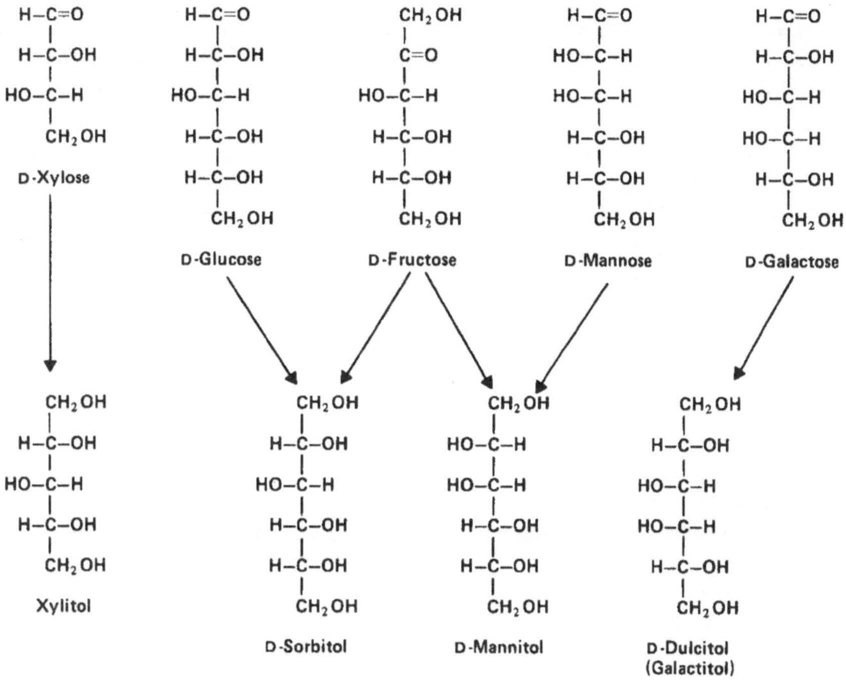


FIGURE 4. Various monosaccharides and the corresponding polyols produced by the reduction of their carbonyl group.

showed that only 16.8% wished to avoid sweets completely. If the diabetic lacks will power in avoiding sweets, and he is directed to avoid their consumption, there is danger that he will consume regular sweets made with sugar. One alternative to this problem is to satisfy the diabetic's sweet tooth with prepared sweets containing noncalorie sweeteners. Chapter 5 covers the application of artificial sweeteners in the preparation of low calorie sweets; therefore, only polyols and fructose will be considered here as sugar substitutes in the diabetic diet. The polyols are reduced carbohydrates produced by reduction of the carbonyl group of carbohydrates as illustrated in Figure 4. The polyols have high water-solubility. Their taste varies from faintly sweet to very sweet (compared to sucrose sweetness). They occur in a variety of plants. Glycerol occurs in fats and other lipids. Erythritol occurs combined as esters in algae and grasses. Ribitol occurs universally in the form of its derivatives – riboflavin, flavin mononucleotide, and flavin adenine dinucleotide. D-sorbitol is one of the most widespread of the naturally occurring polyols. It is found in many fruits, including apples, pears, cherries, plums, and peaches. D-mannitol is widely distributed in plant tissues. This hexitol is present in pumpkin, mushroom, onions, grasses, mistletoe, and lichens. It is found in the mycelia of various fungi. Mannitol occurs in marine aglae, especially brown seaweed, with a seasonal variation in content that can reach over 20% in summer and autumn. Xylitol is found in a number of fruits, vegetables, cereals, mushrooms, lichens, seaweed, and microorganisms. Xylitol is also a normal intermediate product in the glucuronic acid-xylulose cycle in the mammalian metabolism. The following properties justify the inclusion of polyols in the diabetic diet: sweet taste, absorbed slowly (therefore, temporary hyperglycemia observed in the case of sugar is avoided), little effect on blood glucose level, no increase in urinary sugar level, and stabilization of the metabolism. Polyols which are considered of industrial importance include: xylitol, sorbitol, mannitol, maltitol, and lycasin.

*1. Xylitol*

Dehmel et al.[23] studied the absorption of xylitol and other sugars in rats and man. He showed that xylitol is absorbed very slowly in comparison to sucrose. Slow absorption of xylitol is beneficial to diabetic patients because it restricts the rate of xylitol conversion into blood sugar. A slight disadvantage of slow absorption is that a daily consumption of over 40 g xylitol may cause osmotic diarrhea. However, the studies of Bässler and others[24-26] showed that after prolonged feeding, animals usually tolerate a higher dose of xylitol when compared to nonadapted animals. A considerable amount of ingested xylitol was absorbed in the small intestines and ileum, evidently quite faster in adapted rats than in nonadapted rats. The amounts arriving in the lower parts of the intestine were small and still smaller in adapted rats. The long residence of xylitol in lower parts of the intestine causes osmotic effects which explain the observed diarrhea. The different results in adapted rats would count for the observation that diarrhea disappears after prolonged feeding with xylitol.[26]

Xylitol is metabolized by the glucuronic acid-xylulose cycle illustrated in Figure 5. The metabolic reactions generally occur in the liver and kidneys. It is apparent from Figure 5 that through this cycle, xylitol can also be converted to glucose and glycogen. Lang[28] reported that the capacity of man and animals to metabolize xylitol is considerable. Studies in rats with xylitol-U-$^{14}C$ using the continuous i.v. infusion technique showed that about 60% of the dose was oxidized to $CO_2$, and 30 to 35% was retained in the body. Liver accounted for 4 to 6% and the remainder was excreted in the urine. The total xylitol turnover in man has been reported to be in the same range as that of fructose. Based on a 0.4 gram xylitol per kilogram of body weight, the total average turnover in adults was 12 mg/kg/min; 11.3 mg was metabolized and 0.7 mg (6.1% of the dose) was excreted in the urine.[29] The removal of xylitol from the blood and its urinary excretion were found to be the same in adults, children, and even in premature infants, thus suggesting that premature infants are capable of metabolizing xylitol.[28]

Man's ability to metabolize xylitol (to a large extent,) without insulin, its strong antiketogenic effect, and its depressing action on free-fatty acid level in the blood, suggest that xylitol can be useful in the management of diabetes mellitus.[28]

*2. Sorbitol*

Intestinal absorption of sorbitol occurs slowly by means of free-diffusion. At 182, the molecular weight of sorbitol is close to the limit at which diffusion is still possible. As a result, absorption of sorbitol from the intestinal lumen is exceedingly slow. As such, the impelling motive power for resorption is solely the concentration gradient for sorbitol from the intestinal lumen to the blood. Even with the higher concentration of sorbitol in the intestinal lumen, sorbitol can hardly be proven in the blood. It may be concluded from these observations that intestinal resorption of sorbitol can be almost complete since a corresponding concentration gradient always exists.

In the mammalian system, sorbitol is oxidized to fructose by the enzyme sorbitol dehydrogenase which is then metabolized by the fructose-1-phosphate pathway (Figure 6). Depending on the metabolic state of the organism as to the need for glycolysis or gluconeogenesis, sorbitol is converted to pyruvate or glycogen. It has been shown that during the fasting state, during recovery after severe exercise, during cold exposure when glycogen reserves of the body are depleted, or during severe diabetes mellitus, gluconeogenesis proceeds at a higher rate and metabolic products of sorbitol are converted to mainly glucose and glycogen.[29]

Sorbitol has been used as a sweetening agent for diabetes since it was reported in the late 1920s that moderate amounts of sorbitol taken by normal or diabetic subjects caused only a small and insignificant rise in the blood sugar concentration. Adcock and Gray[30] confirmed these observations by feeding uniformly labeled D-U-$^{14}C$ sorbitol and D-U-$^{14}C$ glucose to normal and diabetic subjects. Data showed that in the normal and mildly diabetic subject, over 80% of ingested sorbitol is quickly metabolized to $CO_2$. In a severe diabetic subject requiring insulin, the recovery of labeled $CO_2$ amounted to only 65%, and this lower figure was attributed to increased properties of labeled sorbitol lost in urine. Comparatively, a smaller percentage of glucose was metabolized to $CO_2$ by diabetics and a substantial amount was excreted in the urine. These studies also showed that after 35 g of oral sorbitol, increase in blood sugar was much smaller than after the ingestion of the same quantity of glucose.

In a study carried out by Steinke et al.,[31] sorbitol was added to the diet of 143 out of 205 diabetic children at a summer camp for periods varying from 8 to 48 days. The period of sorbitol administration was preceded and followed by a control period. The remaining 62 diabetic children were treated as a control group who never received

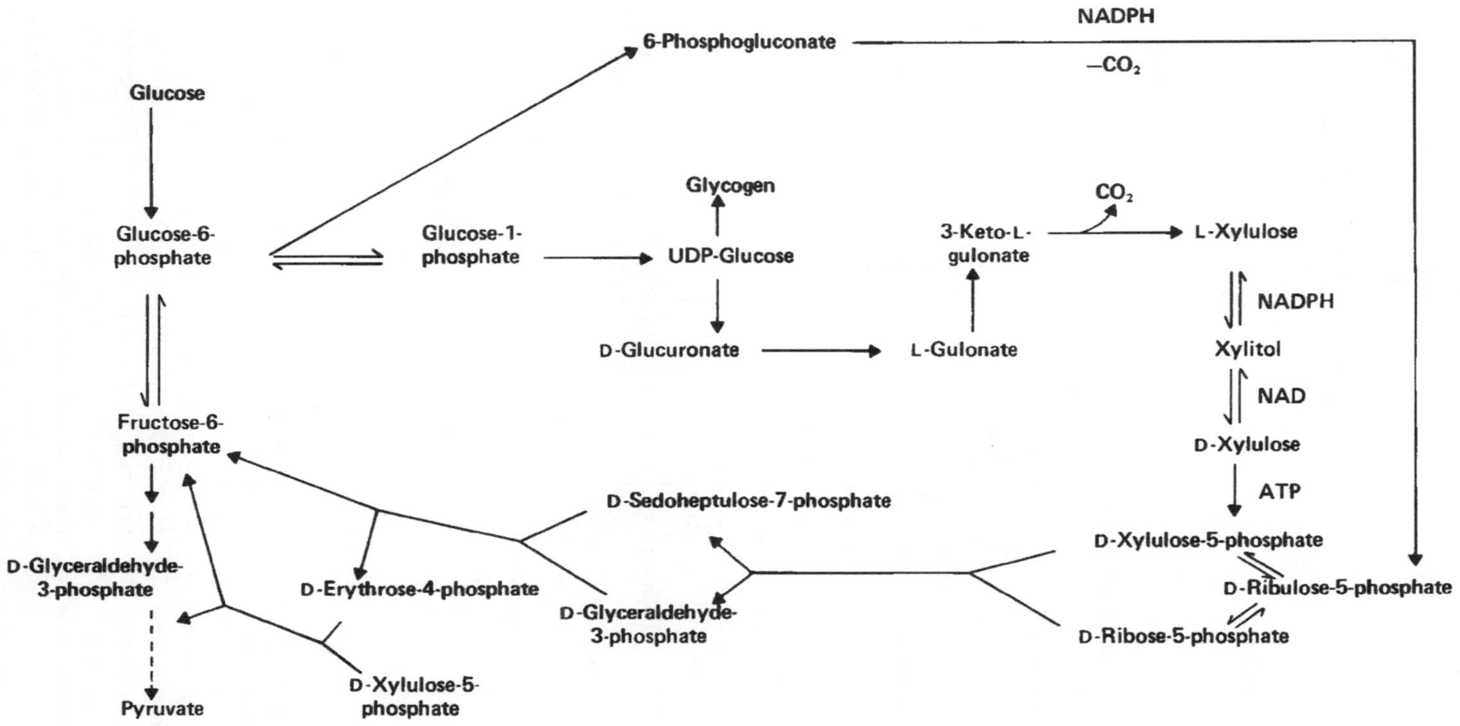


FIGURE 5. Glucuronic acid – xylulose cycle.

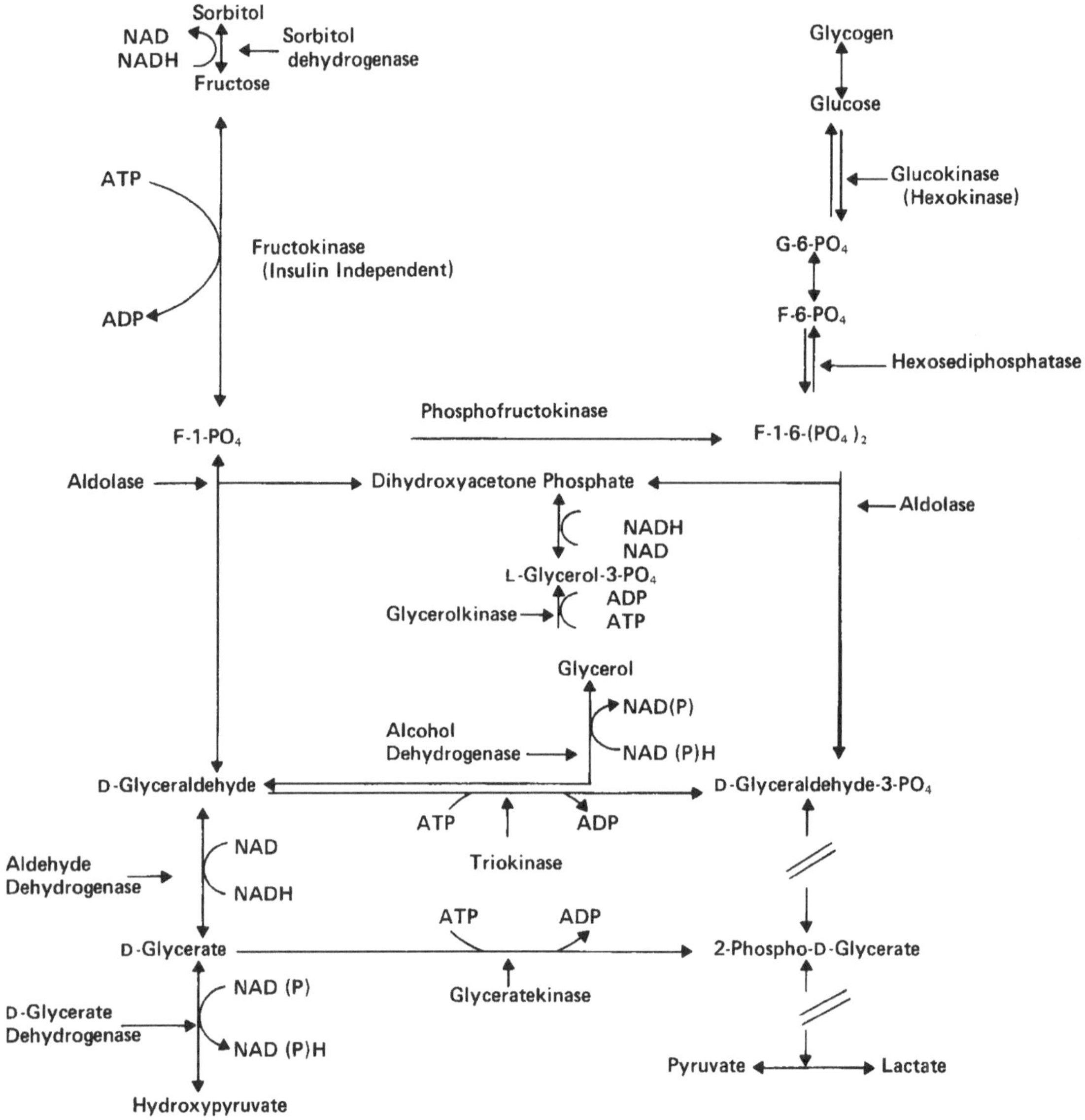


FIGURE 6. Sorbitol metabolism.

sorbitol. Up to 40 g sorbitol daily, given in three daily doses during the meals, was tolerated without diarrhea and without evidence of increased insulin requirement. Glycosuria, daily insulin dose, fasting blood sugar, ketone bodies, and free-fatty acids did not appear to vary significantly as a result of sorbitol incorporation in the diet.[31] Numerous other studies and reports support these findings.[32-39]

Mehnert and Förster[17] recently assessed the usefulness of sorbitol in diabetic diet and concluded that by replacing insulin-requiring carbohydrates with sorbitol, it is possible to improve carbohydrate tolerance to the extent that, in borderline cases, the need for oral hypoglycemic agents will no longer be necessary. In the same way, stabilization of the metabolism by administration of sorbitol, even in the case of patients requiring insulin, is to be expected. For juvenile diabetes having stronger gluconeogenetic activity, there are indications that the use of sorbitol in sensible doses (up to 40 g daily), distributed throughout the day in several individual portions, can prevent deterioration of the metabolism. The slow absorption of sorbitol by the body, from a single dose, results in a slow, continuous release of glucose in the blood for several hours. This property of sorbitol may be effectively used to avoid the disturbing hypoglycemia observed a few hours after insulin injection in severe diabetic subjects, thus stabilizing the metabolism.

*3. Mannitol*

Conflicting reports on the absorption and metabolism of mannitol have been published. Fordtran et al.[40] and Kameda et al.[41] have reported that mannitol remains intact in the intestine if adminstered orally. Stuart et al.[42] reported that it is not metabolized. However, studies by Nasrallah and Iber[43] demonstrate that mannitol is absorbed and metabolized in man. Mannitol-$U^{14}C$ was orally administered in doses of 28 to 100 g. Over 65% mannitol was absorbed, of which a mean 17.5% was recovered in the urine. In these studies, the total amount of mannitol recovered in the stool and urine averaged 50% in 48 hr. The remainder was oxidized, presumably by the liver, to carbon dioxide.

*4. Maltitol*

In man, caloric utilization of maltitol approximates 90%. A portion of maltitol is hydrolyzed in the stomach into its component units, sorbitol and glucose. Sorbitol and glucose are then absorbed and metabolized as discussed earlier. Some of the ingested maltitol is absorbed intact into the small intestine and subsequently split by tissue enzymes. Rennhard and Bianchine[44] recently reported that the principal utilization pathway of maltitol involves fermentation by the large intestinal microflora to volatile fatty acids, which are subsequently absorbed and metabolized.

*5. Lycasin*

Lycasin, composed of hydrogenated analogues of glucose syrup as shown in Table 6, is converted to sorbitol and oligosaccharides in the intestines.

TABLE 6

**Components of Glucose Syrup and Lycasin**

| Glucose syrup | Lycasin |
|---|---|
| Dextrose (D-glucose) | Sorbitol |
| Disaccharides | Hydrogenated disaccharides |
| Trisaccharides | Hydrogenated trisaccharides |
| Polysaccharides | Hydrogenated polysaccharides |

The hydrolyzed products are absorbed and metabolized as sorbitol and glucose by the mechanisms illustrated in Figure 6.

*6. Fructose*

From a comparative investigation of the absorption of sugars and polyols, Dehmel et al.[23] concluded that fructose and disaccharides are absorbed at an intermediate rate, faster than polyols but slower than glucose and galactose. Unlike glucose, absorption of fructose does not require insulin. A small amount of ingested fructose is assimilated and metabolized in cells of the intestinal wall, muscle, and adipose tissue of the periphery. The main cite of fructose metabolism is the liver, where the major portion of ingested fructose is removed immediately after entering the blood stream. Fructose metabolism in the liver proceeds by the fructose-1-phosphate pathways (Figure 6).

The insulin-independent transport of fructose and the initial insulin-independent steps of fructose metabolism offer a significant advantage for fructose over sucrose or glucose in the management of diabetes. Since fructose, (compared to sorbitol) is absorbed at a faster rate, its effectiveness in reducing blood sugar in the diabetic is expected to be lower than sorbitol. Sorbitol, mannitol, xylitol, and lycasin are commercially available and may be used in preparing a variety of dietetic food. General properties of these polyols and their technological applications are listed in Table 7.

## IX. FOODS WITH LOW PROTEIN CONTENT

Low protein foods are recommended for people suffering from chronic uremia, hepatic coma, and acute glomerulonephritis. Under these conditions, the body cannot adequately deal with urea resulting from protein breakdown. To avoid the oxidation of tissues for energy, enough fat and carbohydrates are included in the diet to meet the body's energy requirements. Sufficient protein (20 to 69 g/day) is also included in the diet to replace worn out tissues.[4] Low protein foods can be prepared by selecting ingredients low in protein content or protein-free foods. These foods may be prepared by replacing protein ingredients with modified starches, cellulose derivatives, and gums.

TABLE 7

**Properties and Applications of Polyols**

| Properties | Lycasin | Sorbitol | Mannitol | Xylitol |
|---|---|---|---|---|
| | Product | | | |
| Viscosity of solution | Very viscous | Low | Low | Low |
| Hygroscopicity of solution | + | ++ | ++ | ++ |
| Hygroscopicity of crystalline powder | ++++ | ++ | – | + |
| Crystallization at 20°C (% total solids) | Does not crystallize | 70 | 20 | 63 |
| Sweetness (x sucrose) | 0.75 (type 80/55) | 0.7 | 0.5 or 0.6 | 1 |
| Calorie value | 4 cal/g | 4 cal/g | 2 cal/g | 4 cal/g |
| Insulin sparing effect (Diabetes mellitus) | – | + | + | + |
| Laxative action | – | + | + | + |
| Noncariogenic properties | + | + | + | + |
| **Technological Applications** | | | | |
| Hard-boiled candy | ++ | + | +(?) | – |
| Chewy candy | ++ | +(?) | – | – |
| Chewing gum | + | + | + | + |
| Tablet | – | ++ | + | + |
| Chocolate (plain) | – | + | + | + |
| Gels | + | +(?) | +(?) | +(?) |
| Enrobed goods | – | + | + | – |
| Drageification (candy coating) | – | + | + | ? |

## X. FOODS WITH HIGH PROTEIN CONTENT

High protein foods are used for treatment of certain liver diseases. They are also recommended after surgery, injury, burns, infections, and diseases that lead to increased metabolic rate. One application of high protein foods that has not been seriously considered is in the management of diabetes mellitus. Unlike ingestion of carbohydrates, ingestion of proteins does not stimulate insulin release. Proteins do stimulate the secretion of hormone glucagon by the alpha cells of the pancreas. The metabolic function of glucagon is opposite to that of insulin, i.e., it stimulates the liver phosphorylase, an enzyme necessary in the conversion of glycogen to glucose. Glucagon secretion is stimulated by a drop in the blood sugar level.

## CONCLUSIONS

If the present trend in consumer awareness of nutrition continues, an increased willingness on the part of consumers to pay the increased cost of dietetic foods designed to provide good nutrition and to support the activities which improve nutrition will be evidenced. Optimum nutrition is the key. Overnutrition and undernutrition are both harmful as is evident from the following:

- Protein malnutrition during an infant's first 2 years of life permanently impairs higher mental activities.
- Overnutrition or overfeeding during the first few years of life probably results in an increased number of fat cells. There is now enough evidence that this could greatly increase the chances of obesity in the individual.
- Greater intake of some essential nutrients may also be harmful to the body. A recent report that Vitamin C, in the presence of copper, exhibits a positive mutagenic response to salmonella bacteria is a case in point.

Based on the above, the need for special dietary food designed to correct nutrition imbalance is expected to increase in the coming years. One interesting question which remains unresolved is the relationship between human behavior and diet. It may be noted that most great religions recommend periods of fasting for man to reach deepest spiritual and philosophical insight. These doctrines further suggest a frugal, mostly vegetarian diet for more contemplative intellectual endeavor and a nutrition rich in meat and animal protein for more aggressive individuals.[4,5] It is hoped that nutritionists would assume the challenge to scientifically document the positive relationship that might or might not exist between food composition and higher mental functions. In spite of much work in the field of human nutrition, man is still scratching the surface! It is conceivable that at some future date when enough is learned about human nutrition, dietetic foods would be scientifically designed to suit the mood and the nature of work performed by an individual. This should permit greater satisfaction from our day-to-day activities.

## REFERENCES


1. **Connor, W. E.**, Testimony before the Select Committee on Nutrition and Human Needs, National Nutrition Policy Study – 1974. Part 6. Nutrition and Health, June 21, 1974, U.S. Government Printing Office, Washington, D.C., 1975.
2. **Weir, C. E.**, Benefits from Human Nutrition Research, in Nutrition and Health, with an Evaluation of Nutritional Surveillance in the United States, U.S. Government Printing Office, Washington, D.C., 1975, 67.
3. Code of Federal Regulations 21, Food and Drugs Parts 10 to 199, Definition and Interpretation of Terms, Sec. 125.1, p. 615, U.S. Government Printing Office, Washington, D.C., April 1, 1976.
4. **Bendor, A. E.**, *Nutrition and Dietetic Foods*, Chemical Publishing Company, New York, 1973.
5. **Anon.**, Egg replacer provides caloric reduction, total functionality and convenience, *Food Prod. Dev.*, p. 12, June 1975.
6. **Livingston, G. E.**, The prudent diet: what? why? how?, *Food Technol.*, p. 16, January 1974.
7. **Lachance, P. A.**, Carbohydrates as nutrients, *Food Prod. Dev.*, p. 29, July–August 1973.
8. **Gray, G. M.**, Intestinal digestion and the maldigestion of dietary carbohydrates, *Annu. Rev. Med.*, 222, 391, 1971.
9. **Paige, D. M., Bayless, T. M., Huang, S., and Wexler, R.**, Lactose intolerance and lactose hydrolyzed milk, in *Physiological Effects of Food Carbohydrates*, Jeanes, A. and Hodge, J. Eds., ACS Symposium Series 15, American Chemical Society, Washington, D.C., 1975, 191.
10. **McCormick, R. D.**, A nutritious alternative for the lactose-intolerant consumer, *Food Prod. Dev.* p. 17, June 1976.
11. **Hansen, R. G. and Gitzelmann, R.**, The metabolism of lactose and galactose, in *Physiological Effects of Food Carbohydrates*, Jeanes, A. and Hodge, J., Eds., ACS Symposium Series 15, American Chemical Society, Washington, D.C., 1975.
12. Committee on Medical Aspects of Food Policy, Panel on cariogenic foods: first report, *Br. Dent. J.*, 126, 273, 1969.
13. Scientific Literature Reviews on Generally Recognized as Safe (GRAS) Food Ingredients – Dental caries and carbohydrates PB-241 963/8 ST, National Technical Information Service, U.S. Department of Commerce, Springfield, Va., 1974.
14. **Makinen, K. K.**, The role of sucrose and other sugars in development of dental caries; a review, *Int. Dent. J.*, 22, 363, 1972.
15. **Stephan, R. M.**, Effects of different types of human foods on dental health in experimental animals, *J. Dent. Res.*, 45, 1551, 1966.
16. American Dental Association, Statement of the American Dental Association on Food Advertising before the Federal Trade Commission, October 21, 1976.
17. **Mehnert, H. and Förster, H.**, Oral administration of sorbit as a sugar substitute in the diet for diabetes mellitus, in press, 1977.
18. **Crofford, O. et al.**, Reports to Congress of the National Commission on Diabetes, Public Law 93-354, 93rd Congress, S-2830, July 23, 1974; Findings and Declarations of Purpose Sec 2 (9) (3), 1975, U.S. Government Printing Office, Washington, D.C., cited by Cahill G. F., Etzwiler, D. D., and Freinkel, N., *Diabetes*, 25, 236, 1976.
19. **Spiro, R. G.**, Biochemistry of renal glomerular basement membrane and its alterations in diabetes mellitus, *N. Engl. J. Med.*, 228, 1337, 1973.
20. **Cahill, G. F., Etzwiler, D. D., and Freinkel, N.**, Blood glucose control in diabetes, *Diabetes*, 25, 236, 1976.
21. **Shabo, A. L.**, Report presented at Science Writers' Seminar sponsored by Research to Prevent Blindness, Inc., Reston, Va., May 13, 1976.
22. **Boshell, B. R.**, Diabetes mellitus, current status of therapy, Sherwood Medical Industries, St. Louis, Mo., 197.

23. **Dehmel, K. H., Förster, H., and Mehnert, H.**, Absorption of xylitol, in *Metabolism, Physiology and Clincial Use of Pentoses and Pentitols*, Horecker, B. L., Lang, K., and Tagaki, Y., Eds., Springer-Verlag, Berlin, 1969, 177.
24. **Bässler, K. H., Prellwitz, W., Unbehaun, V., and Lang, K.**, Xylitstoffwechsel beim Menschen. Zur Frage der Eignung von Xylit als Zuckerersatz beim Diabetiker, *Klin. Wochenschr.*, 40, 791, 1962.
25. **Bässler, K. H., Stein, G., and Belzer, W.**, Xylitstoffwechsel und Xylitresorption. Stoffwechseladaptation als Ursache für Resorptionsbeschleunigung, *Biochem. Z.*, 346, 171, 1966.
26. **Bässler, K. H.**, Adaptive processes concerned with absorption and metabolism of xylitol, in *Metabolism, Physiology and Clinical Use of Pentoses and Pentitols*, Horecker, B. L., Lang, K., and Tagaki, Y., Eds., Springer-Verlag, Berlin, 1969, 190.
27. **Touster, O.**, Metabolism and physiological effects of the polyols (alditol), in *Physiological Effects of Food Carbohydrates*, Jeanes, A. and Hodges, J., Eds., ACS Symposium Series 15, American Chemical Society, Washington, D.C., 1975.
28. **Lang, K.**, Utilization of xylitol in animals and man, in *Metabolism, Physiology and Clinical Use of Pentoses and Pentitols*, Horecker, B. L., Lang, K., and Tagaki, Y., Eds., Springer-Verlag, Berlin, 1969, 150.
29. **Pawan, G. L. S.**, Fructose, in *Physiological Effects of Food Carbohydrates* Jeanes, A. and Hodges, J., Eds., ACS Symposium Series 15, American Chemical Society, Washington, D.C., 1975.
30. **Adcock, L. H. and Gray, C. H.**, The metabolism of sorbitol in the human subjects, *Biochem. J.*, 65, 554, 1957.
31. **Steinke, J., Wood, F. C., Donenege, L., Marble, A., and Renold, A. E.**, Evaluation of sorbitol in the diet of diabetic children at camp, *Diabetes*, 10, 218, 1961.
32. **Ellis, F. W. and Krantz, J. C., Jr.**, Sugar alcohols. XXIV. The metabolism of sorbitol in diabetes, *Ann. Intern. Med.*, 18, 792, 1943.
33. **Carr, C. J. and Krantz, J. C., Jr.**, Metabolism of the sugar alcohols and their derivatives, Hexitols. 2. D-Sorbitol, in *Advances in Carbohydrate Chemistry*, Academic Press, New York, 1945, chap. 7.
34. **Wick, A. N., Almen, M. C., and Joseph, L.**, The metabolism of sorbitol, *J. Am. Pharm. Assoc.*, 9, 542, 1951.
35. **Olmsted, W. H.**, The metabolism of mannitol and sorbitol, *Diabetes*, 2, 136, 1953.
36. **Shuman, C. R., Kemp, R. L., Coyne, R., and Wohl, M. G.**, Clinical Use of sorbitol as a sweetening agent in diabetes mellitus, *Am. J. Clin. Nutr.*, 4, 61, 1956.
37. **Felber, J. P., Renold, A. E., and Zahand, G. R.**, The comparative metabolism of glucose, fructose, galactose and sorbitol in normal subjects and disease state, *Mod. Probl. Paediatr.*, 4, 467, 1959.
38. **Hoshi, M.**, Clinical application of sorbitol in patients with diabetes mellitus and in patients with liver disease, *Med. J. Osaka Univ.*, 14(1), 47, 1963.
39. **Mehnert, H.**, Dietary treatment of diabetes mellitus, practical aspects, *Acta Diabetol. Lat.* (Suppl. 1), 9, 530, 1972.
40. **Fordtran, J. S., Rector, F. C., Jr., Ewton, M. F., Soter, N., and Kinney, J.**, Permeability characteristics of human small intestine, *J. Clin. Invest.*, 44, 1935, 1965.
41. **Kamada, H., Abei, T., Nasrallah, S. M., and Iber, F. L.**, Functional and histological injury to intestinal mucosa produced by hypertonicity, *Am. J. Physiol.*, 214, 1090, 1968.
42. **Stuart, F. P., Torres, E., Fletcher, R., Crocker, D., and Moore, F. D.**, Effects of a single repeated and massive mannitol infusion in the dog: structural and functional changes in kidney and brain, *Ann. Surg.*, 172, 190, 1970.
43. **Nasrallah, S. M. and Iber, F. L.**, Mannitol absorption and metabolism in man, *Am. J. Med. Sci.*, 258, 80, 1969.
44. **Rennhard, H. H. and Bianchine, J. R.**, Metabolism and calorie utilization of orally administered maltitol-$^{14}$C in rat, dog, and man, *J. Agric. Food Chem.* 24, 287, 1976.
45. **Altschnl, A. M. and Hornstein, I.**, Foods of the future, *J. Agric. Food Chem.*, 20, 532, 1972.

Chapter 2

# DIETARY LIPIDS IN HEALTH AND DISEASE

**L. J. Vinson**

## TABLE OF CONTENTS



## I. INTRODUCTION

It wasn't too many years ago that dietary fat was best known nutritionally as a rich source of calories: a vehicle for the fat soluble Vitamins A, D, E, and K, for contributing palatability to the diet, and providing a good measure of satiety value. Since World War II, the role of dietary lipids in health and disease has attracted the growing attention of many medical scientists, nutritionists, and epidemiologists. The rising incidence of chronic illnesses that are the major afflictions of our affluent society, viz., atherosclerosis, hypertension, cancer, diabetes, and coronary heart disease (CHD) is the main reason for this. Today CHD is the major public health problem in the U.S. and in many other countries. Dietary fat has definitely been shown to influence serum cholesterol levels in animals and humans. High serum cholesterol levels, in turn, are associated with atherosclerosis, a disease process thought to be initiated by a deposition of cholesterol and other fatty substances in the artery walls. Severe atherosclerosis with its complications – clot formation or thrombosis – is the underlying disease process in most cases of heart attack.

It is the purpose of this chapter to review reports on the effects of commonly encountered dietary lipids in cardiovascular disease. Also, the dietary effects of some uncommon lipids in health and disease will be discussed such as the medium chain triglycerides (MCT), trans acids, and erucic acid. Recent developments linking dietary linoleic acid, the polyunsaturated fatty acid popularly known as EFA, and the prostaglandins will be discussed. The prostaglandins are an exciting, relatively new class of biologically active compounds which apparently act as modulators of cellular functions. For example, they have been shown to have muscle-contracting and blood pressure lowering activities at extremely low concentrations.

When discussing the importance of dietary lipids in the complex interrelations of serum cholesterol, atherosclerosis, and CHD, it must be recognized that other risk factors also identified with CHD exist. Two major risk factors besides diet are hypertension and cigarette smoking. Others are obesity, stress, heredity, diabetes, and life style (sedentary living).

## II. DIETARY FAT AND HIGH SERUM CHOLESTEROL (HYPERCHOLESTEROLEMIA)

In the past 15 years, epidemiologic, clinical, and laboratory studies suggest that dietary lipids wield a strong regulatory action on the levels of circulating lipids as reflected in total serum cholesterol. It has been shown that dietary lipids differ in their effects on serum cholesterol levels.

The commonly encountered dietary lipids can be simply classified as saturated fats, monounsaturated fats, polyunsaturated fats, phospholipids, and nonsaponifiable lipids (Table 1). The amount and type (chain length and degree of saturation) of dietary fatty acids and the amount of dietary cholesterol are known to influence the equilibrium levels of serum cholesterol. In turn, high serum cholesterol levels have been shown to be linked with atherosclerosis. Experimental and clinical studies have established the importance of dietary lipids in the development of high serum cholesterol levels and atherosclerosis. Among the many findings reported in the literature linking dietary fat and serum cholesterol, the following is representative:

TABLE 1

**Types of Dietary Lipids**

| Classification | Lipids |
|---|---|
| Triglycerides | |
| Saturated fat | Lauric, myristic, palmitic, and stearic acids |
| Monounsaturated fat | Oleic and elaidic acids |
| Polyunsaturated fat | Linoleic acid |
| Phospholipids | Lecithin |
| Nonsaponifiable fat | Cholesterol |
| | Plant sterols (sitosterol) |

1. Rabbits fed diets high in saturated fat developed high serum cholesterol levels and a severe atherosclerosis condition. This was not observed in rabbits fed diets high in polyunsaturated fat.[1] Similar findings were obtained with monkeys.[2]

2. Hegsted et al.[3] studied the effects in man of the addition of dietary cholesterol in three diets differing in fat types. One contained coconut oil, high in saturated fat; the second contained olive oil, high in monounsaturated fat; and the third contained safflower oil, high in polyunsaturated fat. Serum cholesterol levels were observed to increase on the coconut oil diet, stabilize on the olive oil diet, and drop significantly in the safflower oil diet. Addition of cholesterol in the form of egg caused a serum cholesterol increase in all three diets at the rate of 5 mg/100 ml or 5mg% serum for each 100 mg cholesterol added. A linear response was obtained to the addition of cholesterol (Figure 1).

3. Keys et al.[4,5] in studies on dietary fat and serum cholesterol, developed multiple regression equations relating change in serum cholesterol to change in the dietary content of saturated (S), monounsaturated (M), and polyunsaturated (P) fatty acids. From some 40 different human diets available for comparison, the equation of best fit in simplified form was

$$\Delta \text{ Chol} = 1.3\ (2S - P) + 1.5Z$$

where $\Delta$ is the change from baseline to the experimental diet; "Chol" is serum cholesterol expressed as mg/100 ml; S and P are percent of total calories per day as saturated and polyun-

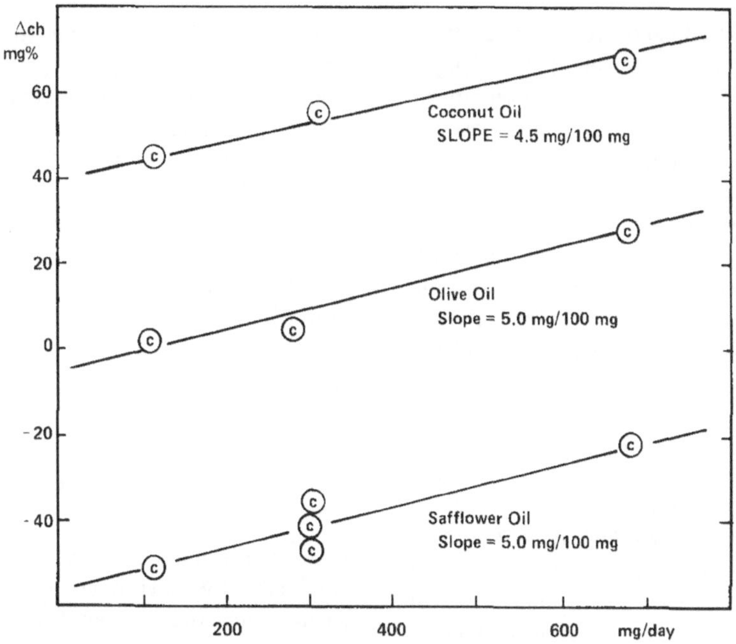


FIGURE 1. Change in serum cholesterol at different levels of cholesterol intake supplied by egg. The diets contained 38% of the calories as fat supplied principally by coconut oil, olive oil, or safflower oil. Taken from McGandy and Hegsted.[7]

saturated fat; and Z is the square root of dietary cholesterol in mg/1000 cal.

4. Hegsted,[3] in a similar study on various diets and effects on serum cholesterol, reported on their overall multiple regression equation simplified as follows:

$$\Delta \text{Chol} = 2.16S - 1.65P + 6.77C - 0.53$$

The symbols in this equation signify the same as in the Keys equation, except for C signifying dietary cholesterol expressed in decigrams per day.

The conclusions drawn from both Keys and Hegsted equations are (1) monounsaturated fatty acids, as found in olive oil or cocoa butter, are essentially neutral, having no effect on serum cholesterol levels; (2) increasing the saturated fatty acid or dietary cholesterol raises serum cholesterol level; and (3) increasing polyunsaturated fatty acids lowers serum cholesterol levels.

Within the class of saturated fatty acids having carbon chain lengths of 12 and higher, the effects on increasing serum cholesterol vary considerably. Both butter fat and coconut oil, known to be very hypercholesterolemic, contain substantial amounts of saturated fatty acids with chain lengths of $C_{14}$ or less. On the other hand, cocoa butter, also high in saturated fat of which over half is stearic acid, ($C_{18}$) the remainder mostly palmitic ($C_{16}$), and about 35% oleic acid, was less hypercholesterolemic than expected according to Hegsted[3] and Keys.[5] These and other findings by Grande[6] strongly suggest that stearic acid like the monounsaturated fatty acid, oleic acid, has no effect upon the serum cholesterol concentration in man. Hegsted has noted[7] that "if stearic acid is indeed substantially less hypercholesterolemic than other saturated fatty acids under dietary conditions, it will be of practical importance in the design of diets aimed at lowering the level of serum cholesterol. Fats such as butter and coconut oil are clearly among the most hypercholesterolemic in the usual diet."

## III. HYPERCHOLESTEROLEMIA, ATHEROSCLEROSIS, AND CHD

A recent American Heart Association report[8] cited statistics that indicate more than 28 million people in the U.S. are afflicted with cardiovascular diseases. In 1972, more than 1 million people died from heart and blood vessel diseases representing

53% of deaths from all causes for that year. Heart attack is the nation's number one killer, claiming more than 600,000 lives a year. About one fourth of the deaths occur in men less than 65 years of age, and an alarming number occur in those in their 40s and 50s. The need for preventive measures is a compelling one.

The production of atherosclerosis in several species of laboratory animals by diet modifications has been demonstrated by many investigators as far back as 1910.[9,10] The development of atherosclerotic lesions is observed by elevating serum cholesterol levels through the feeding of diets high in either cholesterol, animal fat, or both. In its early stages, the atherosclerotic lesion contains large amounts of cholesterol.[11] Atherosclerosis has been shown experimentally to give rise to heart disease in the rat.[12] It has also been demonstrated that monkeys, fed diets high in saturated fat and cholesterol commonly found in American diets, develop atherosclerotic lesions; the severity of the lesion correlates with serum cholesterol level.[2] The lesions regressed when the animals were placed on a low cholesterol, low saturated fat diet which resulted in a drop in serum cholesterol level.

World-wide epidemiologic studies[13-19] have shown a definite correlation between elevated serum cholesterol levels and the incidence of coronary heart disease. A number of workers have undertaken to prove causality in this relationship between cholesterol and heart disease.[20-29] These studies have been of two types: (1) primary prevention studies, wherein individuals with no previous history of coronary heart disease are put on a cholesterol-lowering diet and monitored for the occurrence of coronary heart disease and (2) secondary prevention studies, wherein survivors of a coronary event are placed on a cholesterol-lowering diet and monitored for the occurrence of a repeat myocardial infarction or death ascribed to coronary disease.

One of the longer duration, primary prevention studies was done by the Anti-Coronary Club of New York. After 7 years on a cholesterol-lowering diet, incidence rates of new coronary events were 2.0/1000 person-years for 40- to 49-year-old males and 3.8 for 50 to 59-year-old males compared to rates of 6.4 and 13.3 in control groups of the same age categories.[20] The incidence rates for the control groups on typical American diets high in cholesterol and saturated fat were found to be quite similar to those found for groups of the same ages under observation in other studies and were found to be highly significant ($p < 0.01$). Subjects on the prudent diet, low in cholesterol and saturated fat, showed a drop in serum cholesterol averaging 19 mg/dl (mg %) over the first 4 years of the study compared to the control group. After 10 years of experience, the entire experimental group had a heart disease incidence of 4.3/1000 person-years vs. 10.3 for the control group.[21]

One criticism voiced on the significance of the Anti-Coronary Club study was the failure to make it a double-blind study. The experimental group may have been motivated to take special health precautions by avoiding dietary excesses and adhering to a health care regimen.

A double-blind, primary prevention study was conducted by Dayton et al.[22] Through dietary change, this study achieved a 12.7% reduction in serum cholesterol among residents in a veterans' hospital. Although the incidence of coronary death and myocardial infarction was lower in the experimental group, the total incidence over the study's 8 years was too low to be statistically significant. However, when deaths ascribed to either coronary or cerebral atherosclerosis were added, the difference between experimental and control groups was significant. Death rates for atherosclerotic disease were 14/1000 person-years for the experimental group and 21/1000 person-years for the control group. The death rate was higher than in the Anti-Coronary Club study due to the advanced age of the patients (50 to 89 years).

Primary prevention studies have also been conducted by Turpeinen et al.[23,24] and Stamler.[25] Turpeinen conducted a cross-over design trial in two mental hospitals for two 6-year periods.[23] The regular hospital diet was compared to a special diet including filled (soybean oil) milk instead of regular milk and use of a high polyunsaturated margarine. The special diet was low in cholesterol and saturated fat and contained a high level of polyunsaturated fat. Patients given the special diet showed a drop in serum cholesterol not observed when on the control diet. Analysis of patient adipose tissue showed a significantly higher level of linoleic acid on the special diet when compared to the period on the control diet. For males (pooling the results from both periods), the age-adjusted death rates from CHD were 14.1/1000 person-years on the control diet and

6.6 on the modified diet. The results were highly significant ($p < 0.002$). For females, the pooled death rates were 7.9 for the control periods and 5.2 for the modified diet periods. Results for females did not achieve statistical significance. There were a total of 334 CHD deaths in the study and a duration of 29,217 person-years.[24]

Stamler found that subjects placed on a diet designed to decrease serum cholesterol levels suffered a coronary death rate of only 5/1000 person-years vs. a rate of 20 in a statistically matched group of individuals drawn from participants in other studies.[25]

Secondary prevention studies have been reported by Nelson,[26] Leren,[27] and Bierenbaum.[28] The conclusion of these studies is that fewer recurrences of CHD were associated with lower serum cholesterol and triglyceride levels due to diets low in cholesterol and saturated fat replaced partly with polyunsaturated fat.

In nearly all of the primary and secondary prevention studies, cholesterol levels have been reduced by a combination of methods: reduction of the total fat in the diet, reduction of the cholesterol intake, and increasing the ratio of polyunsaturated to saturated fats in the diet (P/S ratio). All of the cited studies employed diets with higher levels of polyunsaturated fats than those found in the normal American diet or those used in their control diets. In Turpeinen's study, the main diet changes were "replacing (of) ordinary milk by filled milk containing soybean oil instead of milk fat and by replacing butter with a special unsaturated margarine."[23] Similarly, in Dayton's study, the major diet change was substitution of polyunsaturated fats for saturated fat, while the total fat intake remained unchanged (40%) and cholesterol intake remained fairly high[22] (up to seven eggs per week were allowed). This approach was taken because the investigators considered it the least drastic diet change likely to produce a significant cholesterol drop and thus, less likely to be rejected by the subjects.

The use of increased polyunsaturates in the diet to decrease serum cholesterol was first reported by Kinsell and co-workers who investigated the hypocholesterolemic action of polyunsaturated fatty acids under metabolic ward conditions.[29] Subsequent studies supported this finding and led to the previously mentioned empirical equations predicting the effect of dietary saturated fat and cholesterol on serum cholesterol levels by Hegsted et al.[3] and Keys et al.[4] The National Diet-Heart Feasibility Study demonstrated the applicability of these equations to large groups of free-living individuals.[59] This study showed that either increases in the diet P/S ratio or decreases in total dietary fat could lower serum cholesterol and that such changes could be feasibly made in the population at large.

In recent years, authoritative medical and scientific groups, recognizing the seriousness of the CHD problem in the U.S., have recommended changes in the American diet. A joint policy statement of the American Medical Association (AMA) Council on Foods and Nutrition and the National Academy of Sciences-National Research Council NAS–NRC Food and Nutrition Board declared that "the evidence now available is sufficient to discourage further temporizng with this major health problem."[30] The statement also indicated that the cholesterol level in most people can be lowered "most practicably by partial replacement of the dietary sources of saturated fat with sources of unsaturated fat, especially those rich in polyunsaturated fatty acids, and by a reduction in the consumption of foods rich in cholesterol."

A report published in 1970 entitled, "Primary Prevention of Atherosclerotic Disease[31]", suggested that reducing serum cholesterol by as little as 5% based on epidemiologic studies, would yield a 12% decrease in CHD incidence. This small drop in serum cholesterol can be readily achieved by modifying dietary fat intake by replacing part of the saturated fat with polyunsaturated fat and reducing the dietary intake of cholesterol. Associated with this 5% drop in serum cholesterol would be an estimated decrease in CHD deaths of over 72,000/year in the U.S.

## IV. THE NEED TO MODIFY THE AMERICAN DIET

In their 1970 report,[31] the Inter-Society Commission for Heart Disease Resources stated that urgent public health decisions to modify the American diet must be made. Furthermore, the report says we cannot wait for the complete solution to the problem of atherogenesis or for a complete understanding of the way multiple-risk factors interact to produce atherosclerosis and CHD. The Commission recommends that the medical profession and the government supported by the general public encourage the food industry to produce and market fat-modified foods low in saturated fat and cholesterol. A strategy for primary prevention of premature CHD events is recommended as long-term national policy for the U.S. Implementation of this strategy by providing funds and manpower to accomplish certain objectives involves diet changes preventing or controlling high serum cholesterol, obesity, hypertension, and diabetes; elimination of cigarette smoking; and pharmacologic control of elevated blood pressure.

Because differences of opinion still exist regarding the beneficial effect of various types of diet change (particularly fat modification) on premature CHD in the U.S., the government is sponsoring clinical studies to gain additional substantive evidence on the importance of diet in preventing this disease.

One such study, the Multiple Risk Factor Intervention Trial (MR. FIT), assesses effectiveness of measures to reduce elevated blood cholesterol, high blood pressure, and cigarette smoking in preventing first heart attacks and reducing death rates from CHD. This government study is supported by the National Heart and Lung Institute and the National Institute of Health (NIH). NIH is one of six agencies comprising the Public Health Service.

Twenty clinical centers throughout the U.S. are taking part in the MR. FIT Program. Screening procedures in the first stage identify and enroll 600 CHD-prone volunteers per clinical center, achieving a total of 12,000 needed for the study.

Volunteers selected from men aged 35 to 57, free of clinical CHD, are above average CHD risks due to various combinations of elevated blood cholesterol, high blood pressure, or cigarette smoking. Eligible and willing participants are divided into two groups. Half are referred to their personal physicians who are told about the screening test results. These participants return to the clinical center for cardiovascular examinations each year over the 6-year test period. The other half participate in a special intervention program designed to reduce blood cholesterol levels by diet modifications, lower elevated blood pressure, and eliminated cigarette smoking. Intervention in this group involves changing some living and health habits and requires a strong educational component. The staff of each clinical center includes a cardiologist for diagnostic evaluations, nutritionists to help in dietary changes, and behavioral scientists to aid in smoking cessation.

The MR. FIT Program and other clinical studies emphasizing the role of diet in heart disease will take up to 10 years for completion and determination of results. Regardless of this, and based on the large amount of evidence already amassed, the prevailing medical opinion is to act now against CHD, the number one public health problem.

The Inter-Society Commission for Heart Disease Resources and other major medical groups recommend the following actions in their position papers on CHD:

1. Calorie intake be adjusted to achieve and maintain optimal weight
2. Reduction of dietary cholesterol to 300 mg or less per day
3. Substantial reduction of dietary saturated fats by partial replacement with polyunsaturated fat to provide a 1:1 ratio (P/S = 1)
4. Reduction in simple refined sugars contained in candy, soft drinks, and other sweets and greater use of complex natural carbohydrates such as vegetables, fruits, and cereals
5. Avoidance of excessive salt in the diet

To keep the consumer better informed on the nutritional value of food products (particularly relating to the dietary factors associated with CHD), the Food and Drug Administration (FDA) issued regulations for nutrition labeling in 1973, seeking to encourage food manufacturers to provide such information on nutrients.

The FDA's goal is to provide uniform nutrient information on different foods. This information provides the following consumer benefits:

1. Awareness of key nutrients needed for good nutrition and health
2. Recognition of the specific nutrients present in individual foods
3. Comparison of the nutritive values and relative costs of different foods
4. Simplified selections for the daily diet supplying the recommended amounts of vital nutrients

Medical findings linking diet and heart disease and easing of stringent government regulations for health-oriented foods have encouraged food manufacturers to develop and market modified food products having the palatability, attractiveness, and economics of comparable standard food items. These food products lack such health shortcomings as high saturated fat and/or cholesterol.

In recent years, innovative products low in saturated fat and cholesterol and usually containing a high level of desired polyunsaturated fat were introduced to the food market. Examples of these products include the following.

**High polyunsaturated margarines** – Margarines containing a very high level of polyunsaturated fat are currently available. These margarines are low in saturated fat and free of cholesterol. The use of delta lactones and other flavoring compounds found in butter give them all the desirable flavor characteristics, making them the products of choice because of their health benefits. Clinical tests demonstrate that high polyunsaturated margarines, as part of prudent dieting or simply as replacements for butter in diets of institutionalized free-living adults, resulted in a significant drop in serum cholesterol (see Table 2).

**Egg substitutes** – Several liquid products now available on the market make reasonably good omelets and can replace eggs in a wide range of recipes. These egg substitutes do not contain yolk, the source of undesirable cholesterol and saturated fat in eggs. Used regularly as part of a prudent diet, they should contribute significantly to lower blood cholesterol levels.

**Cheese products** – These include soft cheeses like cottage cheese, ricotta, and mozzarella prepared from low-fat skim milk. Healthwise, they are preferable to hard cheeses like cheddar which are high in saturated fat and cholesterol. Cheese substitutes, claiming low or no cholesterol content, are now appearing in the marketplace. It is anticipated that technology will improve in the next few years and provide a variety of cheese-like products with the flavor acceptability of conventional cheese counterparts, but without high amounts of saturated fat and cholesterol.

**Meat substitutes** – The appearance of meat-like products prepared with textured vegetable protein (soybean) have appeared on the market. Products simulating both flavor and appearance of sausage patties, link sausages, and ham patties are gaining consumer acceptance. Textured vegetable protein is widely applied as an extender for meat, fish, and seafood products. Its use contributes to lowered cholesterol and lowered saturated fat in the daily diet.

The need to alert Americans about cardiovascular perils of the average American diet typically high in saturated fat and cholesterol, prompted the Inter-Society Commission for Heart Disease Resources to make several recommendations.[31] The Commission advocates that the medical profession the government, and the general public encourage the food industry to make available leaner processed meats, dairy products, frozen desserts, and baked goods reduced in saturated fats, cholesterol, calories, visible fats, and oils. These foods and also margarines, shorten-

TABLE 2

**Hypocholesterolemic Effect of Diets Containing Polyunsaturated Margarine**

| Site | Population type | Diets | Test design | Subjects | Reduction in average serum cholesterol (mg %) |
|---|---|---|---|---|---|
| Edgewater, N.J. | Free | Butter vs. high PUFA[a] margarine | Cross-over, 6 weeks | 356 | 6.0 |
| Edgewater, N.J. | Free | Butter vs. high PUFA margarine | Cross-over, 3 weeks | 100 | 7.5 |
| Concord, N.H. School | Free | Typical American diet vs. prudent diet with high PUFA margarine | Linear, school year and vacation | 192 | 22.0 |
| Essex County, N.J. | Institutionalized | Butter vs. high PUFA margarine | Linear, 3 weeks | 267 | 15.8 |
| New York, N.Y. | | Butter vs. high PUFA margarine | Cross-over, 3 weeks | 218 | 10.2 |
| Montclair, N.J. | Institutionalized | Typical American diet vs. prudent diet with high PUFA margarine | Linear, 4 weeks | 115 | 32.2 |

[a]Polyunsaturated fatty acid.

ings, mayonnaises, salad dressings, and oils should be low in saturated fat, cholesterol content, and high in polyunsaturated fat. The Commission further recommended that to effect the required changes in dietary habits, it is essential that the entire community be actively involved through a comprehensive and sustained public and professional education program. This will deepen understanding and appreciation of the need for primary prevention and will inform the public and health professions on methods of food selection and preparation consistent with sound nutritional practices.

## V. MEDIUM CHAIN TRIGLYCERIDES (MCT)

Triglycerides containing medium chain fatty acids principally with chain lengths of $C_8$ and $C_{10}$ (Figure 2) differ from those containing the longer chain fatty acids (myristic, palmitic, and stearic) with respect to their digestion, absorption, transport, and metabolism. Long chain triglycerides are known to undergo intestinal hydrolysis to monoglycerides, fatty acids, and glycerol by pancreatic lipase before absorption; are reesterified to triglycerides within the intestinal mucosa; and are then transported as chylomicra via the lymphatics. In contrast, MCT are far more readily digestible even in the absence of bile. They are quickly absorbed intact and undergo hydrolysis within the intestinal mucosa. They are then transported not through the lymphatics but via the portal vein to the liver. MCT, unlike the long chain triglycerides, is readily metabolized to $CO_2$. Very little is incorporated into hepatic lipids compared to long-chain triglycerides. Because of its absorption ease, calories in MCT are available to maintain positive caloric balance and body weight in those individuals unable to effectively digest and absorb conventional dietary fats.

MCT is particularly valuable in the malabsorption syndrome arising from pancreatic lipase deficiency and bile salts as reported by Hashim et al.[32]

MCT has also been found valuable in preterm infant feeding.[33] The neonate as well as the older infant is characterized by a temporary inability to absorb fatty acids; a physiological steatorrhea often develops. In such instances, MCT in milk formulas is readily absorbed by the newborn.

MCT is commercially prepared with the $C_8$ and $C_{10}$ fatty acids separated from hydrolyzed coconut oil. Ironically, coconut oil, high in stearic acid and identified as a risk factor in atherosclerosis, serves as a source of MCT not implicated in atherosclerosis. In this connection, milk fat triglycerides contain about 13% of medium chain fatty acids ($C_{6-10}$).

## VI. TRANS FATTY ACIDS

Monounsaturated and polyunsaturated fats are comprised of fatty acids such as oleic acid and linoleic acid, respectively, which have double bonds in the cis configuration. Trans acids are found commonly. Their structural distinctions are illustrated in Figure 3.

Trans acids exist naturally. For example, animal fats contain small amounts of trans fats, with 5 to 10% trans acids found in butter. Many vegetable oils also contain traces of trans acids. However,

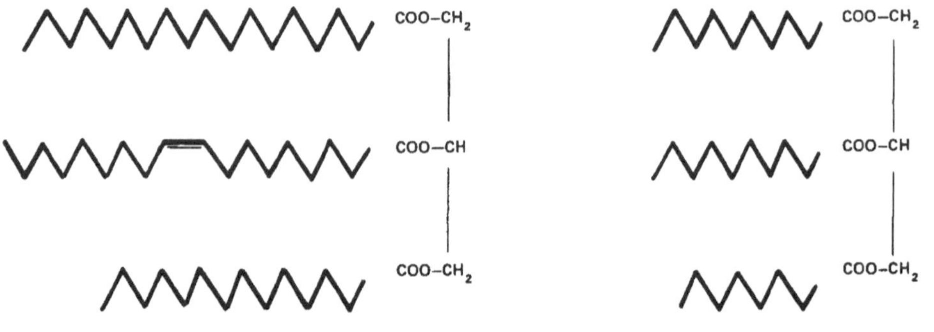


Glyceride of Long Chain Fatty Acids

Glyceride of Medium Chain Fatty Acids

FIGURE 2. Structure of long and medium chain triglycerides.

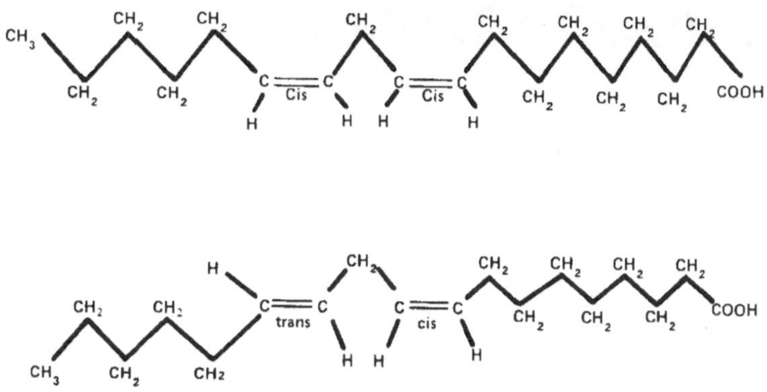


FIGURE 3. Structure of cis, cis linoleic acid, and cis, trans linoleic acid.

when such oils are "hardened" to raise their melting point, the cis unsaturated fatty acid partially transforms to the trans form occurring with the desired partial hydrogenation. Margarines, many salad oils, and other food products contain partially hydrogenated oils to provide desirable plasticity and keepability. The fat composition of such products may contain up to 25 to 30% trans acids.

Extensive scientific literature exists on the metabolism of trans fats. Humans and animals utilize trans fatty acids as nutrient sources, being readily metabolized to $CO_2$ and water. Trans oleic acid (elaidic) is catabolized at a rate similar to that of palmitic and oleic acids. The trans isomers of linoleic acid are also metabolized to $CO_2$ at a rate somewhat superior to cis, cis linoleic acid. In one study, rats were fed a diet high in a vegetable fat containing 35% trans acids. Growth, reproduction, and lactation through 47 generations were normal and compared favorably with control rats.[34]

In rat studies investigating the influence of trans acids on serum cholesterol, results indicate that trans acids do not exhibit hypercholesterolemic effect. Human studies also determined that serum cholesterol levels show no practical relationship to the quantity of trans acids in the diet. In view of these findings on trans acids, a report by Kummerow in 1974,[35] suggesting that trans acids were implicated in the development of hypercholesterolemia and atherosclerosis in swine, comes somewhat as a surprise.

The practical significance of Kummerow's results are questionable because of certain deficiencies in the study's experimental design. For example, the basal diet was criticized for its subnormal protein and essential fatty acid levels. Studies previously reported in the literature indicate that swine, fed fat high in trans acids but containing adequate amounts of essential fatty acids, did not exhibit atherosclerosis.[36]

## VII. ERUCIC ACID (RAPESEED OIL)

Rapeseed crops contribute substantially to the vegetable oil market. They are grown successfully in Canada, Europe, and even in subtropical areas. Oils extracted from these seeds have a long history of consumption by man. Production of rapeseed oil on a worldwide basis places its tonnage in fifth place as an edible oil after soybean, sunflower seed, ground nut, and cottonseed oil. Most rapeseed oil is used in salad oils, cooking oils, margarine, and shortenings.

Rapeseed oil has a characteristic high level of erucic acid (cis docosenoic acid) in its triglycerides. Erucic acid is a monounsaturated fatty acid of the n-9 fatty acid family which also includes oleic acid (Figure 4).

Thomasson and Boldingh[37] showed that erucic acid is responsible for a growth-retarding action observed in diets high in rapeseed oil. Erucic acid possesses peculiar physiological effects. Several investigators reported acute infiltration of fat into heart muscle in several animal species (rat, mouse, and duckling). With prolonged feeding, the fatty infiltration gradually disappears and is replaced by an irreversible fibrosis. Apparently the heart muscle cell, dependent on fatty acids for its energy needs, fails to obtain an adequate supply because high erucic acid levels in the diet partially block fatty acid transport into the cell.[38,39]

Although no adverse effects from erucic acid have been reported in man, animal findings have

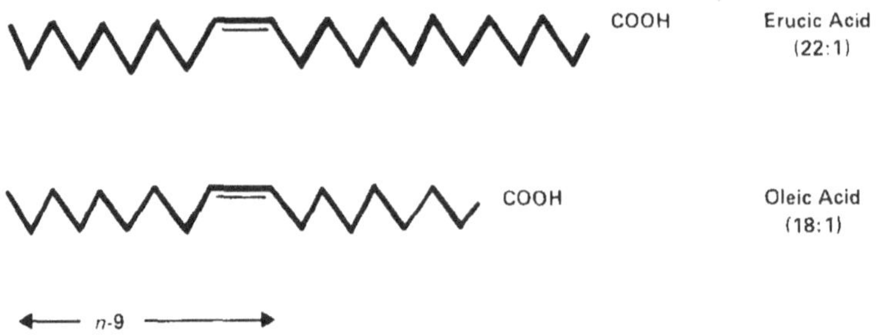


FIGURE 4. A structural comparison of erucic acid and oleic acid, both members of the *n*-9 fatty acid series.

prompted government action to restrict its use in foods. Three years ago, Canada, switched its production of rapeseed crop from the conventional 25 to 30% erucic acid variety to a new variety (Canbra) with virtually no erucic acid. Sweden, France, and Germany are expected to soon follow a similar policy.

## VIII. DIETARY FAT AND ARTERIAL THROMBOSIS

The final insult in CHD is the formation of a thrombus blocking the blood supply to the heart muscle. This results in myocardial infarction or heart attack. Formation of a stable thrombus is the result of a complex process involving, in the initial stages, circulating blood platelets which under certain conditions will adhere to subendothelial tissue and form aggregates. These platelet clusters release substances that promote further aggregation and activate the clotting process. Thus the thrombus formation comprises two main processes: formation of platelet aggregates and fibrin formation.

Dietary fats influence arterial thrombosis. Hornstra and Vendelmans[40] demonstrated that rats fed diets high in linoleic acid showed a much lower tendency to develop arterial thrombosis than rats on diets prepared with saturated fat. The results of studies done on rats indicate:

1. Saturated fatty acids promote arterial thrombosis formation (blood platelet aggregation time is shortened).
2. Oleic acid and elaidic acid are neutral to arterial thrombus formation.
3. Linoleic acid has specific antithrombotic effect (blood platelet aggregation time is extended).

There is no significant relationship between plasma cholesterol content and blood platelet tendency to aggregate.

Findings in rats were confirmed in humans as reported by Hornstra et al.[41] Studies indicated that subjects fed a polyunsaturated diet exhibited significantly lower blood platelet aggregation tendency than those fed a saturated diet. The subjects were part of a long-term clinical study[23] indicating that incidence of CHD was reduced on a polyunsaturated diet. In this study conducted by Turpeinen in Finland, patients fed a high linoleic acid diet (12 cal %) showed a significant drop in blood cholesterol and a decreased tendency for blood platelet aggregation and thrombosis compared to the control patients on a normal Finnish diet containing 4 cal % of linoleic acid. Fleischman et al.[42] also reported on the beneficial effect of a polyunsaturated diet on the in vivo platelet function in man. Their data showed that increasing dietary linoleate for 2 weeks in a group of 66 institutionalized residents was associated with a significant increase in blood platelet stability reflected in doubling of the aggregation time. Decreasing dietary linoleate in the following 2 weeks resulted in a reverse effect. These subjects participated in the Essex County location clinical study mentioned previously (Table 2) and indicated a significant drop in serum cholesterol when on the diet containing a high polyunsaturated margarine.

It is believed that dietary linoleic acid plays several roles in combating premature CHD. Linoleic acid controls serum cholesterol levels and minimizes the tendency of blood platelets to aggregate which is the initial stage in atherosclerosis and thrombus formation. Thomasson[43] postulated that linoleic acid is a precursor of prostaglandin $E_1$, the active principle in reducing the tendency of blood platelets to aggregate. Van Dorp et al.[44] have shown that feeding linoleic acid to rabbits and rats caused increased prostaglandin synthesis. Recent studies demonstrate other important functions of dietary linoleic acid:

1. Increased dietary linoleic acid results in a significant increase in coronary blood flow and enhances myocardial contractility.[45,46]

2. A high linoleic acid diet may be beneficial in cases of hypertension. Rat studies demonstrating antihypertensive activity of dietary linoleic acid fed to rats made hypertensive on a high NaCl intake[60] suggest this.

3. A linoleic acid diet results in improvement of hyperlipidemia and abnormal serum insulin levels of obese diabetic patients.[47]

Thus, dietary linoleic acid is an important nutritional factor in minimizing or preventing several pathological conditions linked with cardiovascular disease. They are hypercholesterolemia, arterial thrombosis, hypertension, and maturity onset diabetes.

Prostaglandins synthesized by the body from dietary linoleic acid are a group of biologically active substances characterized by the basic structure of a fatty acid with a carbon ring (cyclopentane) and two hydrocarbon chains attached to two adjacent carbon atoms. They contain 20 carbon atoms, a hydroxyl group at $C_{15}$, and a trans double bond between $C_{13}$ and $C_{14}$. Two prostaglandins, $PGE_1$ and $PGE_2$, are shown in Figure 5 with their precursor fatty acids: linoleic and arachidonic acid. $PGE_1$ is a potent inhibitor of platelet aggregation while $PGE_2$ shows no such activity.

Both $PGE_1$ and $PGE_2$ have hypotensive effects, antagonize the hypertensive properties of adrenalin and angiotension, are potent inhibitors of catecholamine-induced free fatty acid release from adipose tissue, increase natriuresis, decrease the influence of both sympathetic and parasympathetic nerve stimulation on myocardial tissue, and have a very potent antiarrhythmic and inotropic activity on the myocardium.

Prostaglandins are generally not considered hormones. They are synthesized by practically all tissue cells. They are primarily considered to act as modulators of cellular functions. Prostaglandins are formed *in situ* and subsequently eliminated via the blood circulation system having a half-life of only a few minutes. They are uniquely characterized by their ubiquity (being found in most tissue cells), high potency at extremely low concentrations, and a remarkable diversity of physiological functions. Prostaglandins appear to participate in kidney function, intestinal peristaltics, inflammation, skin function, and the central nervous system. They can regulate both carbohydrate metabolism and fat mobilization.

## IX. LECITHIN AND ATHEROSCLEROSIS

Atherosclerotic lesions contain cholesterol

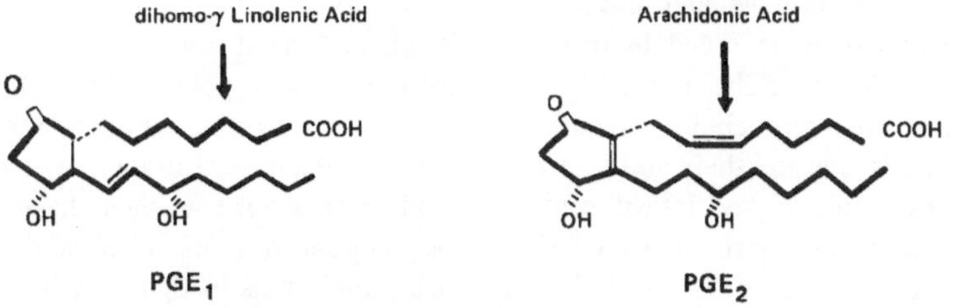


FIGURE 5. Structures of linoleic acid, precursors, and prostaglandins $PGE_1$ and $PGE_2$.

which is virtually a water-insoluble compound. Reversal of this condition requires removal in the form of fatty acid esters. The polyunsaturated fatty acid esters of cholesterol are least sclerogenic and are readily removed from tissue deposits. Thus, their formation should be beneficial in the prevention and resolution of atherosclerotic plaques according to Krumdieck and Butterworth.[48]

There is evidence that the enzyme lecithin cholesterol acyltransferase (LCAT) is responsible for cholesterol esterification. This enzyme catalyzes the transfer of a fatty acid acyl residue from the 2-position of a lecithin to the hydroxyl group of cholesterol. Individuals who lack this enzyme develop hyperlipidemia and atherosclerosis at an early age.[49]

Not all lecithins will serve equally well as substrates for the LCAT enzyme. The effective lecithins are those bearing linoleic acid in position 2 to promote the synthesis of polyunsaturated cholesterol esters which favor the removal of cholesterol deposits. Soybean lecithin possesses a high quantity of linoleic acid whereas egg lecithin contains predominantly saturated fatty acids. Adams et al.[50] demonstrated that polyunsaturated lecithin accelerated cholesterol resorption much more effectively than relatively saturated lecithins. Armstrong and Megun's findings in monkeys support the concept that regression in atheromatous lesions can be achieved with polyunsaturated lecithin.[51]

## X. SITOSTEROL AND HYPERCHOLESTEROLEMIA

Sitosterol is a plant sterol possessing a structure identical to cholesterol. The difference lies in the side chain of the cyclopentenophenanthrene ring. Campesterol, stigmasterol, and sitosterol are the most common plant sterols encountered in edible oils and the human diet. Many investigators report that plant sterols fed to animals and humans inhibit blood cholesterol increase noted in those on diets containing high levels of cholesterol. It is widely believed that the hypocholesterolemic effect of plant sterols is due to their ability to inhibit cholesterol absorption in the intestine.[52] Other mechanisms probably exist in view of studies that indicate a lowering of blood cholesterol values in man and animals when sitosterol is fed in a diet free of cholesterol.[53]

## XI. MULTIPLE SCLEROSIS

In recent years, a dietary link with multiple sclerosis has been reported.[54] Several investigators have drawn attention to the relationship between high incidence of mulitple sclerosis and diets high in saturated fatty acid in some parts of the world. Swank and Bourdillon[55] recommend reducing the total fat and replacing part of the saturated fat with polyunsaturated fat in diets of multiple sclerosis patients. Improvement in the patient's functional condition with attendant-reduced relapsed rate was noted. Sunflower seeds or corn oils are suggested supplements to the patient's diets. This is indicated in view of findings that the linoleate levels in multiple sclerosis patients was lower than in the controls, while that of oleate was higher.[54] In such patients and control subjects, feeding supplements of sunflower seed resulted in a significant rise in linoleic acid levels of the blood. When supplementation was stopped, there was more rapid drop in blood levels of multiple sclerosis patients than in normal subjects. For some unexplained reason, the rate of loss or of utilization of linoleate is abnormally high in patients with multiple sclerosis. However, it is not warranted to conclude at this time that multiple sclerosis is due to a dietary deficiency of polyunsaturated fatty acids.

## XII. DIETARY FAT AND CANCER

Evidence exists that dietary lipids may contribute to the development of human cancer. Dietary excesses have been associated with lower gastro-intestinal tract and endocrinal organ cancer. Epidemiologically, a correlation between fat consumption and colon cancer has been reported by Wynder and Shigematsu.[56] Cited is the rareness of such cancer among Japanese in Japan but it is observed to increase among Japanese immigrating to Hawaii or California, suggesting that differences in dietary habits are responsible. The Japanese diet is characterized by a small percentage of fat calories, mostly polyunsaturated, in contrast to 40 calories % fat or more in the American diet, nearly half of which is saturated. Wynder reported that studies on stool specimens have shown that high risk populations for colon cancer differ from low risk populations in that they have a relatively high content of anaerobic bacteria as well as bile acid and cholesterol metabolites in their feces, reflecting presumably a dietary intake high in fat and cholesterol.

High fat and high cholesterol diets have been linked also epidemiologically with breast cancer. Lea[57] reported that a highly significant correlation exists between consumption of fats and oils in different countries and death rates from neoplasms of the breast ($p < 0.001$). This association of high dietary fat levels and cholesterol with cancer suggests epidemiological association between CHD and several forms of cancer. Using the death rates from 24 countries, Lea found positive correlations between coronary heart disease and (1) breast neoplasms ($p < 0.001$), (2) leukemia (over the age of 55) $0.02 > p > 0.01$, and (3) rectal neoplasms ($0.05 > p > 0.02$). Positive correlation was found between death rates from CHD and the consumption of fats and oils ($0.02 > p > 0.01$). Lea concludes that his analysis of fats and oils statistics suggests that cholesterogenic diets linked with CHD may also favor the growth of certain neoplasms.

## CONCLUSIONS

This review highlighted recent research progress on the importance of dietary lipids in health and disease. Certainly dietary lipids cannot be considered just a calorie source and a vehicle for fat-soluble vitamins. The essential fatty acid, linoleic acid, an important factor in cardiovascular health, precursor of prostaglandins, a class of compounds that modulate metabolic and physiologic pathways. Considerable research on prostaglandins continues in many medical and industrial laboratories in the U.S. and Europe. It is anticipated that new biological functions will be uncovered. The greatest promise of the prostaglandin studies now in progress lies in the areas of cell regulations, pathology, and therapeutics.

The role of diet remains somewhat controversial despite overwhelming experimental, clinical, and epidemiological evidence linking dietary fat, hypercholesterolemia, atherosclerosis, and CHD. Failure to generally agree on the role of diet in CHD prevention stems primarily from the multifactorial nature of cardiovascular disease. A recent report[58] by a joint Working Party of the Royal College of Physicians of London and the British Cardiac Society succinctly defines the problem by stating that the causes of CHD are largely environmental and rooted in the modern, affluent way of life. Regarding diet, the Working Party makes recommendations that the whole community modify dietary fat intake by reducing saturated fat and partially replacing it with polyunsaturated fats.

Practical considerations on the importance of dietary lipids in health and disease with focus principally on CHD prevention can be best summarized by quoting from the position statement of the British Working Party:[58]

> This Working Party has tried to steer a course between the academic and the pragmatic views of CHD prevention, between the nihilism of those awaiting final scientific proof before taking action and the enthusiasm of those determined to eliminate every risk factor from all individuals. We consider that the present size of the CHD problem in this country and the small effect of medical and surgical treatment on the mortality rate from CHD justify attempts to prevent the disease we cannot cure. Research and controlled trials must continue but it now seems an appropriate time for energetic action based on a balance of probabilities. The measures we recommend have a reasonable hope of conferring some benefits, and none of them has a cost that approaches the cost of inaction. The degree of energy with which these measures should be applied must depend on the risk in the individual and on the choices the individual decides to make. His ability to make decisions regarding his life pattern and the risk of future disease will depend on his understanding of the concept of risk.

# REFERENCES


1. Roberts, J. C. and Strauss, R., Eds., *Comparative Atherosclerosis,* Harper & Row, New York, 1965.
2. Armstrong, M. L., Warner, E. I., and Connor, W. E., Regression of coronary atherosclerosis in rhesus monkey, *Circ. Res.,* 27, 59, 1970.
3. Hegsted, D. M., McGandy, R. B., Myers, M. L., and Stare, F. J., Quantitative effects of dietary fat on serum cholesterol in man, *Am. J. Clin. Nutr.,* 17, 281, 1965.
4. Keys, A., Anderson, J. T., and Grande, F., Serum cholesterol response to changes in the diet. II. The effect of cholesterol in the diet, *Metabolism,* 14, 759, 1965.
5. Keys, A., Anderson, J. T., and Grande, F., Serum cholesterol response to changes in the diet. IV. Particular saturated fatty acids in the diet, *Metabolism,* 14, 776, 1965.
6. Grande, F., Anderson, J. T., and Keys, A., Comparison of effects of palmitic and stearic acids in the diet on serum cholesterol in man, *Am. J. Clin. Nutr.,* 23, 1184, 1970.
7. McGandy, R. B. and Hegsted, D. M., Quantitative effects of dietary fat and cholesterol on serum cholesterol in man, in, *The Role of Fats in Human Nutrition,* Vergroesen, A. J., Ed., Academic Press, New York, 1975, 211.
8. Heart Facts 1975, The American Heart Association, New York.
9. Katz, L. N. and Stamler, J., *Experimental Atherosclerosis,* Lea & Febiger, Philadelphia, 1958.
10. Katz, L. N., Stamler, J., and Pick, R., *Nutrition and Atherosclerosis,* Lea & Febiger, Philadelphia, 1958, Chap. 5.
11. National Heart and Lung Institute Task Force on Atherosclerosis, Atherosclerosis, Vol. 2, Department of Health, Education and Welfare Publ. No. (National Institutes of Health) 72-119, Washington, D.C., 1971.
12. Hartcroft, W. S. and Thomas, W. A., Pathological lesions related to disturbance of fat and cholesterol metabolism in man, *JAMA,* 164, 1899, 1957.
13. Keys, A., Coronary heart disease in seven countries, *Circulation,* (Suppl.), 41, 42, 1970.
14. Stamler, J., Diet and atherosclerotic disease, *J. Am. Diet Assoc.,* 34, 701, 1958.
15. Jolliffe, N. and Archer, M., Statistical associations between international coronary heart disease death rates and certain environmental factors, *J. Chronic Dis.,* 9, 936, 1959.
16. Miller, D. C., Trulson, M. F., and McCann, M. B., Diet, blood lipids and health of Italian men in Boston, *Ann. Intern. Med.,* 128, 907, 1971.
17. Tyroler, H. A., Heyden, S., Bartel, A., Cassel, J., Cornoni, J. C., Hanes, C. G., and Kleinbaum, D., Blood pressure and cholesterol as coronary heart disease risk factors, *Arch. Intern. Med.,* 128, 907, 1971.
18. Carlson, L. A. and Bottiger, C. E., Ischaemic heart disease in relation to fasting values of plasma triglycerides and cholesterol, *Lancet,* 1, 865, 1972.
19. Kannell, W. B., Dauber, T. R., Kagan, A., Revotaskie, N., and Stokes, J., Factors of risk in the development of coronary heart disease – six years follow-up experience. The Framingham Study, *Ann. Intern. Med.,* 55, 33, 1961.
20. Christakis, G., Rinzler, S. H., Archer, M., Winslow, G., Jamkel, S., Stephenson, J., Friedman, G., Fein, H., Kraus, A., and James, G., The Anti-Coronary Club. A dietary approach to the prevention of coronary heart disease – a seven year report, *Am. J. Public Health,* 56, 299, 1966.
21. Christakis, G. and Rinzler, S. H., Therapy; diet, in *Atherosclerosis,* Schottler, F. G. and Boyd, G. S., Eds., Elsevier, New York, 1969, Chap. 10.
22. Dayton, S., Pearce, M. L., Hoshimoto, S., Dixon, W. J., and Tomiyasu, U., A controlled clinical trial of a diet high in unsaturated fat, *Am. Heart Assoc. Monogr.,* 25, 1969.
23. Turpeinen, O., Miettinen, M., Karvonen, M. J., Roine, P., Pekkarinen, M., Lehtosuo, E. J., and Aliverta, P., Dietary prevention of coronary heart disease; long term experiment, *Am. J. Clin. Nutr.,* 21, 255, 1968.
24. Miettinen, M., Turpeinen, O., Karvonen, M. J., Esosno, R., and Paavilainen, E., Effect of cholesterol-lowering diet on mortality from coronary heart disease and other causes, *Lancet,* 2, 7782, 1972.
25. Stamler, J., Acute myocardial infarction; progression primary prevention, *Br. Heart J.,* 33, (Suppl. 145) 1971.
26. Nelson, A. M., Treatment of atherosclerosis by diet, *Med. Welt,* 1965, 2036, 1965.
27. Leren, P., The Oslo diet heart study – eleven year report, *Circulation,* 42, 935, 1970.
28. Bierenbaum, M. L., Fleischman, A. L., Green, D. P., Rachelson, R. I., Hayton, T., Watson, P. B., and Caldwell, A. B., The five-year experience of modified fat diets on younger men with coronary heart disease, *Circulation,* 42, 943, 1970.
29. Kinsell, L. W., Partridge, J., Boling, L., Morgan, S., and Michaels, G., Dietary modification of serum cholesterol and phospholipid levels, *J. Clin. Endocrinol.,* 12, 909, 1952.
30. American Medical Association Council on Foods and Nutrition and Food and Nutrition Board of the National Academy of Sciences – National Research Council, Dietary Coronary Heart Disease, *JAMA,* 222, 1647, 1972.
31. Inter-Society Commission for Heart Disease Resources, Primary Prevention of Atherosclerotic Diseases, *Circulation,* 42, (Suppl. A), 55, 1970.
32. Hashim, S. A., Studies of medium chain fatty acid transport in portal blood, in *Medium Chain Triglycerides,* Senior, J. R., Ed., University of Pennsylvania Press, Philadelphia, 1968.
33. Tantibhedhyangkul, P. and Hashim, S. A., Medium-chain triglycerides feedings in premature infants: Effects on fat and nitrogen absorption, *Pediatrics,* 55, 359, 1975.

34. **Alfin-Slater, R. B., Wells, P., Aftergood, L.**, Dietary fat composition and tocopherol requirement. IV. Safety of polyunsaturated fats, *J. Am. Oil Chem. Soc.*, 50, 479, 1973.
35. **Kummerow, F. A.**, Lipids in Atherosclerosis at Symposium on Nutritional Perspectives and Atherosclerosis, paper presented at the American Oil Chemists Meeting, New Orleans, La., May 14, 1974.
36. **Roswell, H. C., Downie, H. G., and Mustard, J. F.**, *Can. Med. Assoc. J.*, 79, 647, 1958.
37. **Thomasson, H. J. and Boldingh, J.**, The biological value of fats and oils: II. The growth retarding substance in rapeseed oil, *J. Nutri.*, 56, 469, 1955.
38. **Abdellatif, A. M. M.**, Cardiopathogenic effects of dietary rapeseed oil, *Nutr. Rev.*, 30, 2, 1972.
39. **Ten Hoor, F., van de Graaf, H. M., and Vergroesen, A. J.**, Effects of dietary erucic acid and linoleic acid on myocardial function in rats, in, *Myocardial Metabolism, Recent Advances in Studies on Cardiac Structure and Metabolism,* Dhalla, N. S., Ed., University Park Press, Baltimore, 1973.
40. **Hornstra, G. and Vendelmans, A.**, Introduction of experimental arterial occlusive thrombi in rats, *Atherosclerosis,* 17, 369, 1973.
41. **Hornstra, G., Lewis, B., Chart, A., Turpeinen, O., Karvonen, M. J., and Vergroesen, A. J.**, Influence of dietary fats on platelet function in man, *Lancet,* 1, 1155, 1973.
42. **Fleischman, A. L., Simpkins, H., and Bierenbaum, M.**, Beneficial effect of increased dietary linoleate upon *in vivo* platelet function in man, *Circulation,* 48 (Suppl. 4) 55, 1973.
43. **Thomasson, H. J.**, Prostaglandins and cardiovascular diseases, *Nutr. Rev.*, 271, 67, 1969.
44. **Van Dorp, D. A., Beerthuis, R. K., Nugteren, D. H., and Vonkeman, H.**, Enzymatic conversion of all-cis polyunsaturated fatty acids into prostaglandins, *Nature,* 203, 839, 1964.
45. **Vergroesen, A. J., de Deckere, E. A. H., ten Hoor, F., Hornstra, G., and Houtsmiller, U. M. T.**, Physiological functions of essential fatty acids, in *Essential Fatty Acids,* Hawkins, W. W., Ed., The Nutrition Society of Canada, 1975, 5.
46. **ten Hoor, F. and Vergroesen, A. J.**, Prostaglandins and the heart, *J. Mol. Cell. Cardiol.*, 536, 1975.
47. **Houtsmuller, U. M. T.**, The role of fat in the treatment of diabetes mellitus, in *The Role of Fats in Nutrition,* Vergroesen, A. J., Ed., Academic Press, London, 1975.
48. **Krumdieck, C. and Butterworth, C. E.**, Ascorbate-cholesterol-lecithin interactions: factors of potential importance in the pathogenesis of atherosclerosis, *Am. J. Clin. Nutr.*, 27, 866, 1974.
49. **Norum, K. R., Borsting, S., and Grundt, I.**, Familial lecithin: Cholesterol acyltranferase deficiency, *Acta Med. Scand.*, 188, 323, 1970.
50. **Adams, C. W. M., Abdulla, Y. H., and Morgan, R. S.**, Modification of aortic atheroma and fatty liver in cholesterol fed rabbits by intravenous injection of saturated and polyunsaturated lecithins, *J. Pathol. Bacteriol.*, 94, 77, 1967.
51. **Armstrong, M. L. and Megun, M. B.**, Lipid depletion in atheromatous coronary arteries in Rhesus monkeys, *Circ. Res.*, 30, 675, 1972.
52. **Hernandex, H. H., Peterson, D. W., Charkoff, I. L., and Dauben, W. G.**, Absorption of cholesterol-4-$C^{14}$ in rats fed mixed soybean sterols and β-sitosterol, *Proc. Soc. Exp. Biol. Med.*, 83, 498, 1953.
53. **Beveridge, J. M. R., Connel, W. F., Mayer, G. A., and Haust, H. L.**, Plant sterols, degree of unsaturation, and hypocholesterolemic action of certain fats, *Can. J. Biochem. Physiol.*, 36, 895, 1958.
54. Editorial, Fats and multiple sclerosis, *Br. Med. J.*, June 5, 1971.
55. **Swank, R. L. and Bourdillon, R. B.**, Multiple sclerosis: assessment of treatment with a modified low-fat diet, *J. Nerv. Ment. Dis.*, 131, 468, 1960.
56. **Wynder, E. L. and Shigematsu, T.**, Environmental factors of cancer of the colon and rectum, *Cancer,* 20, 1520, 1967.
57. **Lea, A. J.**, Dietary factors associated with death rates from certain neoplasms in man, *Lancet,* 2, 332, 1966.
58. Joint Working Party of the Royal College of Physicians, London and the British Cardiac Society, Prevention of coronary heart disease, *J. R. Coll. Physicians London,* 10, 214, 1976.
59. National Diet-Heart Study Research Group, The National Diet-Heart Study Final Report, *Am. Heart Assoc. Monogr.*, 18, 1968.
60. **Triebe, G., Block, H. U.**, Förster, W., Über das Blutdruckverhalten Kochsalz-belasteter Ratten bei unterschiedlichen Linolsäurebehalt des Futters, *Acta Med. Biol. Germanica,* 1976, in press.

Chapter 3
# LOW CALORIE BULKING AGENTS

**J. J. Beereboom**

## TABLE OF CONTENTS



## I. INTRODUCTION

The world may encounter the specter of future food shortage. Already, hundreds of thousands of people exist at the edge of famine; thus, it is relevant to establish the need for low calorie foods before discussing use of bulking agents for preparation of these foods. In many countries, industrialized society provides a plentiful food supply and demands a sedentary life style. Consequently, obesity has become a common and serious disorder, affecting from 15 to 40% of the adult population, depending upon the criteria for obesity used.[1] Obesity is associated with excess mortality from cardiovascular disease, hypertension, respiratory illness, and diabetes.[2] Diabetes, particularly of the maturity-onset type, is increasing and is now the third highest illness causing death in the U.S. A strong association between obesity and diabetes exists. Fully 90% of the 5 million diabetic Americans were overweight when the disease was first diagnosed. It is also well established that the clinical and chemical manifestations of maturity-onset diabetes may regress following weight reduction.[3] Certainly, the consumer seems well aware of the hazards of obesity. In a recent nationally conducted survey, 46% of the individuals questioned described themselves as "overweight," and fully two thirds were attempting to lose weight by dieting and/or exercise.

While the simplistic, popular prescription, "eat less – exercise more" works satisfactorily with highly motivated individuals, it is inadequate for the majority. Many people drift from one diet fad to another; they obtain temporary weight loss from each regimen but soon relapse, and their weight returns to the same or an even higher level – a pattern once referred to as the "rhythm method of girth control."[14]

## II. WEIGHT CONTROL TECHNIQUES

Two relatively recent approaches to weight reduction demonstrate sufficient promise and deserve special mention. The first, used largely for severely obese patients under proper medical supervision, involves partial starvation diets with lean body mass maintained by small protein amounts.[5] This protein-sparing diet shows very good results when massive weight reduction is needed. The second approach, based on the discoveries of Van Itallie and others, recognizes that while the normal individual sets his food consumption pattern in response to physiological-oriented hunger signals transmitted to a "satiety center," the obese individual responds to a whole series of nonphysiological cues leading to overconsumption.[6] Recognition of this has led to a series of behavioral modification approaches to break undesirable eating habits, particularly those of the mildly obese or overweight individual. Good weight-reduction results. Adequate weight maintenance over the long term has been reported for such a behavioral modification program.[7] The utility of low calorie foods to reinforce desirable behavioral patterns should be appreciable, but very few studies have been done establishing actual benefit.

To be successful, any dietary regimen must be sufficiently attractive so the consumer will stay with it in the weight-reduction phase and continue it to maintain ideal body weight. A reasonable objective for the moderately obese individual in a weight-reduction phase might be a caloric deficit of 500 cal/day, allowing a weight loss approaching 1 lb/week. Since the excess weight generally accumulated over a long period, it would appear prudent for these individuals to lose it slowly, insuring that the diet remains nutritionally adequate. The weakness of such a program is that many obese individuals, when put on a calorically-restricted diet, temporarily decrease energy expenditure and consequently reduce gradually causing the process to be discouragingly slow.[8]

What ingredients should be restricted in the diet? It is interesting to note that, for reasons unrelated to obesity, various authorities have suggested reductions in dietary saturated fats,[9] refined carbohydrate,[10] and protein.[11] Moderate reductions in these ingredients could in themselves lead to appreciable caloric reductions. For the obese and the diabetic, nutritionally adequate diets have been constructed around the following approximate distribution:

| Ingredient | Calories (%) | Intake at 800 cal/day (g) | Intake at 2000 cal/day (g) |
|---|---|---|---|
| Protein | 20 | 40 | 100 |
| Fat | 35 | 31 | 78 |
| Complex carbohydrates | 35 | 70 | 175 |
| Simple sugars | 10 | 20 | 50 |

Some diabetologists recommend even higher carbohydrate levels and lower fat levels than shown in the above table, but it is difficult to provide a palatable diet having such low fat levels. This distribution, used as an extreme reducing diet of 800 cal/day, is obviously quite restrictive and somewhat protein deficient. However, as a maintenance diet of 2000 cal/day, it allows an attractive, variable food list. Two major problems in maintaining such a diet for extended periods are limiting portion sizes and severe restriction of those socially oriented as they cannot indulge in dessert and snack foods to which they are accustomed. Within this area especially, low calorie foods as dietary aids can have a most positive impact. Availability of good quality, low calorie foods could provide the overweight consumer with enough additional motivation to endure a diet plan or a weight maintenance program.

Consumer activists claim that low calorie foods are worthless since they make no significant nutritional contribution to the diet in terms of essential ingredient intake. Such reasoning may be appropriate for evaluating diets for beef cattle in a feed lot, but it ignores the critical esthetic and social features of human food consumption. Providing these features without desirable weight gain would be a positive benefit – at least in the same sense that modern man can have the enjoyment of sex without the consequence of procreation.

## III. DIET FOODS

Caloric reduction can be accomplished by reduced portion size (thinly sliced bread) and occasionally by omission of an ingredient as in dietetic soft drinks. However, in most foods, reduction of the amount of ingredients sufficient to cause significant caloric reduction leads to loss of appearance, texture, and mouth-feel readily detected by the consumer. Thickeners and emulsifiers often improve the viscosity and general appearance of many reduced bulk foods,[12] but usually the consumer notes a decrease in expected mouth-feel that limits acceptability. One could also speculate about the possible effects of these reduced bulk foods in terms of a "satiety factor."

The desired bulk and mouth-feel of most socially oriented desserts are provided by carbohydrate and fat components. Low calorie food of good quality can only be made with bulking agents having physical properties permitting replacement of the bulk and functional properties of the usual fat or carbohydrate. In addition to this quality restraint, there is also the cost problem. It is safe to assume that such bulking agents will be more expensive than the ingredients they replace. Though desserts and snacks are luxury items (nutritionally unnecessary), there is a price limit beyond which the consumer justifiably will not pay. The food scientist invariably faces a compromise situation in these requirements when designing new low calorie food formulations.

## IV. DETERMINATION OF CALORIC UTILIZATION

Direct measurement of caloric utilization in the human subject (performed by Atwater and his associates around the turn of the century) has proven accurate. For all intents and purposes, the values of many food ingredients measured by Atwater are presently used. However, the procedure is truly a herculean task for the investigator and requires elaborate equipment. It demands a level of volunteer cooperation extremely difficult to obtain today. Reference 13 provides an excellent review of the Atwater investigations.

A common procedure for estimating caloric utilization is the method originated by Rice et al., who compared weight gains of weanling rats on a basal diet to those of rats on the same diet supplemented with the test ingredient.[14] For efficiently utilized materials like the fats originally investigated by Rice and co-workers, significant results can be obtained by testing a few animals over a short test period. However, when the procedure is used to evaluate material with low digestibility, the scatter of data often increases so much that large numbers of animals must be used on various control regimens before statistically significant results occur. One frequent complication occurring with low digestible material is cecal enlargement caused by osmotic water retention,[15] which, or course, can modify body weight.

In vitro enzymatic hydrolysis measurements are frequently used to estimate caloric utilization. Numerous references in the literature claim that, since a particular polysaccharide is not easily attacked by the usual digestive enzymes, it must have a low-caloric utilization value. This reasoning totally ignores the lower gut fermentation processes occuring constantly as part of normal digestion. These will be discussed in greater detail later. However, the method is useful for accumulating preliminary screening data.

Perhaps the most reliable method for establishing low-caloric utilization of ingested material is to recover it from fecal samples. Unfortunately, quantitative measurement of material from feces, particularly polysaccharides, can be exceedingly difficult and applicable only to material that passes through the gastrointestinal tract unchanged.

Metabolism studies conducted with appropriately radiolabeled samples provide an accurate measurement of caloric utilization. The method is limited only by availability of suitable tagged samples and the equipment necessary for collecting appropriate samples and measuring radioactivity in breath, serum, urine, and fecal samples. Dosages with a $^{14}C$ specific activity as high as 50 $\mu$Ci can be safely administered to humans and provide sufficient sensitivity and precise data. The method is unsuitable for human experiments in rare instances requiring repeated dosing, since the accumulative radioactive dose may be hazardous. In this regard, it should be noted that there has been a recent proposal to take advantage of the difference in natural abundance of $^{13}C$ occurring in certain forms of starch.[16] This difference is large enough to be measured by mass spectrometry; thus the starch, the glucose derived from the starch, or conversion products from either would possess a suitable label that could be measured and repeatedly administered to humans.

Though there are several examples of reducing food's caloric value by substitution of a high calorie ingredient with one of lower caloric value, there are only two obvious ways that a material can function as a low calorie bulking agent by decreased utilization. In the first, a rare situation, the material absorbed from the gastrointestinal tract is rapidly excreted via the kidneys before the liver has a chance to utilize it in catabolic reactions. The variable, but low, caloric utilization reported for mannitol is caused partially by its urinary excretion pattern.[17] In the second, a far more common situation, the material passes through the gastrointestinal tract without being absorbed. However, even highly insoluble material, which at first glance appears to be completely nonutilizable, is seldom completely inert. Numerous examples of this phenomenon will be discussed in detail below. Suffice it to say that the operative mechanism for utilization in such cases does not usually involve the normal acid-digestive enzyme process, but instead depends on microbial fermentation by organisms resident in the lower gut. This well-recognized but poorly understood process accounts for an appreciable portion of the available calories in the normal ruminant diet and for a smaller but significant caloric intake portion in man and other nonruminants. The vast array of hydrolytic enzymes available from the large variety of microorganisms in the gastrointestinal tract make it unlikely that any bulking agents discussed below can pass through the gut without undergoing some change.

Normally, the microorganisms utilize the enzymatic hydrolysis products as carbon sources for growth and excrete products, particularly volatile fatty acids absorbed by the mammalian host and utilized for energy, and in the general carbon pool cycles. Unfortunately, the microbial fermentation causes production of appreciable amounts of carbon dioxide, hydrogen, and methane. These gases (combined with any non-absorbed, volatile, fatty acids) are largely responsible for flatulence and stool softening (due to osmotic changes) frequently observed when a poorly digestible material is consumed in large quantities. The effect can range from a mild, barely perceptible increase in gas production associated with normal consumption of foods such as beans or cabbage to outright but temporary diarrhea occurring in lactase-deficient individuals when they consume excessive quantities of milk products.[18]

Therefore, with many bulking agents, caloric utilization and human tolerance are functions of the gut microflora's fermentation capability. A soluble, easily fermented substrate such as sorbitol will have a high caloric utilization value, but will be tolerated poorly in high doses. An insoluble, poorly fermented substrate such as cellulose should have insignificant caloric utilization and will be well tolerated except for some slight stool softening due to its bulk effect. Exceptions to this general rule are certain hydrocolloids that are only slightly utilized, but bind sufficient water thus producing an appreciable laxative effect even at very low concentrations.

The remainder of this chapter discusses various bulking agents that have been used or proposed as replacements for fat or for carbohydrate in flour, starch, and sugar.

## V. BULKING AGENTS FOR FAT

The simplest approach towards caloric reduction in fat-based foods is reduction of fat content. The ready consumer acceptance of low fat milk products is a good example, and low fat margarines produced by special emulsification techniques have also had considerable success, particularly in certain European countries.[19] Similar approaches are being used for dietetic mayonnaise, salad dressings, and ice creams with varying success. Within this group of products, consumer acceptance has been limited.

A subtler approach has been taken in the use of medium chain triglycerides where the principal fatty acids are those with eight and ten carbons. These acids (hence the corresponding triglycerides made from them) have somewhat lower caloric values than the usual long chain fatty acids because of the higher oxygen to carbon ratio of the medium chain acids. The medium chain triglycerides are also reported to be less efficiently converted to cholesterol after absorption.[20] The medium chain triglycerides are liquid at room temperature, so there are formulation problems to be overcome. However, they have been successfully used to produce a calorie reduced margarine in Europe.[21]

The Proctor and Gamble Laboratories report some potentially interesting low calorie fat replacements.[22] Sucrose and a number of polyols fully esterified with oleic acid were increasingly resistant to hydrolysis by pancreatic and a non-specific lipase as the number of ester functions increased. This work recalls an earlier finding that fully esterified, hindered alcohols such as neopentylglycol and pentaerythritol are not utilized efficiently as calorie sources.[23] The Proctor and Gamble compounds containing six (sorbitol) and eight (sucrose) ester groups were not hydrolyzed.[24] This resistance to hydrolysis accounts for the lack of pseudofat absorption in the rat gut.[25] The most interesting of these compounds, a sucrose polyester that is a mixture of the hexa-, hepta-, and octa-esters of long chain (largely C-18) fatty acids, was shown to decrease cholesterol absorption in the rat.[26] A brief communication[27] suggests that the sucrose polyester has excellent organoleptic characteristics as a fat substitute.

Since over 98% of a 50-g ingested dose could be recovered from human stools, the noncaloric nature of these sucrose polyesters appears well established. Under metabolic ward conditions, normal volunteers who ingested the sucrose polyester for 10 days showed a significant decrease in serum cholesterol and low density lipoprotein. A small but significant decrease in serum vitamin A values occurred. Unfortunately, patients with familial hypercholesterolemia, a group in most need of treatment to lower serum cholesterol, did not show any significant change in serum lipid values under the test conditions.[27] Evidence from the animal and human experiments suggests that the nonabsorbed sucrose polyester binds dietary cholesterol and prevents its absorption. Whether the sucrose polyesters will play a role as noncaloric, hypocholesteremic food fat replacements will probably depend upon the economics of producing such products on a large scale.

Polyoxyethylene fatty acid esters and polyglycerol esters represent a slightly different approach to the preparation of fat replacements with lower caloric value. The first class of compounds, typified by polyoxyethylene stearate, is more useful as an emulsifying agent than as a fat replacement per se. The caloric contribution depends almost entirely on the stearate portion since most of the polyoxyethylenediol part of the molecule, liberated by gut lipase hydrolysis of the ester, is excreted unchanged. The caloric value of the ester is thus only about 4.2 cal/g, an appreciable reduction from the usual 9 cal/g for fats.[28]

The polyglycerol esters, which represent a group of products with a wide range of solubilities and physical properties, have been cleared for use as emulsifiers in foods by the Food and Drug Administration (FDA). Some of the esters have excellent oil- and fat-like characteristics.[20] Though caloric utilization of these materials has not been measured, the values should be quite low since the polyglycerol portion of the ester should not be absorbed and metabolized, a situation analogous to the polyoxyethylenediols.

In the late 1950s, a number of fatty acid esters of amylose and monoglyceride derivatives of dibasic acids such as adipic and fumaric acids were developed at the United States. Department of Agriculture (USDA) Southern Regional Research Laboratory. These were proposed as possible fat replacements for use in surface coatings or slab dressings in foods.[29] Caloric utilization of many of these materials was low as indicated by the usual rodent digestibility procedures.[30] However, considering the difficulties in using digestibility studies to evaluate materials with low-caloric utilization, the low-caloric characteristics of these esters remain to be proved.

A glycan obtained from baker's yeast that is probably a mixture of 1-6 and 1-3 glucans and 1-6 mannan has also been approved for use in foods.[31] Though it is not a low calorie substance, it is reported to have good fat-sparing effects in many foods.

## VI. BULKING AGENTS FOR CARBOHYDRATES

Two types of bulking agents for carbohydrates are needed: (1) a soluble material that can replace sucrose and other simple carbohydrates in food and (2) an insoluble material capable of replacing flour or starch. The structurally more complex insoluble type of carbohydrate bulking agent will be considered first. The most outstanding characteristic of this group is the necessity (which is seldom met) for physical properties duplicating the rheological properties of flour and starch. In most instances, only partial substitution of flour and starch is possible. This limits the caloric reduction achievable in the food. Another characteristic of known insoluble bulking agents is that only limited caloric utilization data are available for most of them.

The so-called dietary fibers are the most structurally complicated low-caloric bulking agents. Based on rather weak epidemiological evidence, various authors have made positive health claims for dietary fibers in the alleviation or prevention of diverticular disease, colonic cancer, varicose veins, hemorrhoids, venous thrombosis, atherosclerosis, gallstones, appendicitis, diabetes, and obesity. Many of these claims are currently under investigation in appropriate clinical studies. Evidence for a beneficial effect in diverticular disease, which strikes a large proportion of the population over the age of forty, for example, appears to be relatively good. However, until the necessary studies have been performed, there will be disagreement over the value of dietary fiber. Part of the current confusion is due to lack of agreement over what actually constitutes fiber. It can be obtained from a variety of plant sources. At this time, there is no general agreement on the analytical procedures that should be used to measure dietary fiber in foods, although there is agreement that the traditional analysis for crude fiber is inappropriate for dietary fiber determination. Recently, the entire subject of health claims and analytical problems of dietary fiber has been excellently reviewed by Spiller and Amen.[32]

It is generally accepted that dietary fiber is at least composed of hemicelluloses, cellulose, and lignins in varying proportions depending upon the plant source and the extraction procedure. Many authors of review articles assume that dietary fiber passes through the human gastrointestinal tract without change. This position completely ignores the classical studies performed by Olmsted in 1936.[33] Using the admittedly crude separation and analytical techniques available at the time, he measured the lignin, cellulose, and hemicellulose content of various dietary fibers in substances ranging from cabbage to agar-agar. A small group of volunteers was given appropriate diets, and the collected feces were analyzed for the three components. Olmsted concluded that lignin was hardly utilized, but cellulose disappeared to the extent of about 45% and hemicellulose to approximately 70%. Changes in the amount of volatile fatty acids in the stools was also measured. From these data, Olmsted proposed a utilization mechanism for these dietary fiber components involving lower gut microbial fermentation and associated volatile fatty acid production. The amount of acids produced correlated well with the increase in stool weight and change in stool consistency. Olmsted concluded that the laxative effect of dietary fibers was due primarily to the degree of microbial utilization of components rather than to any bulk effect of the fibers. The study lacks adequate statistical control for use in making judgments by modern standards on the caloric utilization of dietary fiber, but it does provide fascinating reading for those interested in how discoveries can be made, forgotten, and reinvestigated a generation later.

Many of the proponents of dietary fiber have recommended regular consumption of isolated wheat bran as means of increasing fiber intake, but many people find the substance unpalatable. Recently, bran and other fibers have been incorporated into processed foods, particularly breads and cereals. Though these foods significantly increase fiber intake, the fiber's physical properties frequently cause sufficient textural problems so that only modest caloric reductions of the food are possible, and some palatability problems remain.

Another point regarding dietary fiber consumption deserves comment. Although certain investigators recommend increased consumption to decrease the cholesterol and bile salt absorption, others express concern about the effect dietary fiber might have on absorption of essential dietary ingredients, particularly minerals. Weak evidence suggests that dietary fiber decreases the gastrointestinal absorption of minerals such as calcium, magnesium, and zinc.[34] Whether this is due to the well-understood and thoroughly investigated influence of the fiber phytate content or to changes in transit time and stool consistency caused by fiber cannot be determined by presently available data. While it remains to be proven that dietary fiber can have a significant effect on mineral utilization, this problem is bound to influence regulatory decisions on fiber use in foods.

The most thoroughly studied, insoluble bulking agent is isolated cellulose. Metabolism studies using $^{14}C$-labeled cellulose have conclusively demonstrated that the rat is capable of utilizing 30 to 50% of the material.[35,36] The mechanism of utilization is undoubtedly microbial cellulose fermentation in the rat's cecum. For man, conflicting statements exist in the literature con-

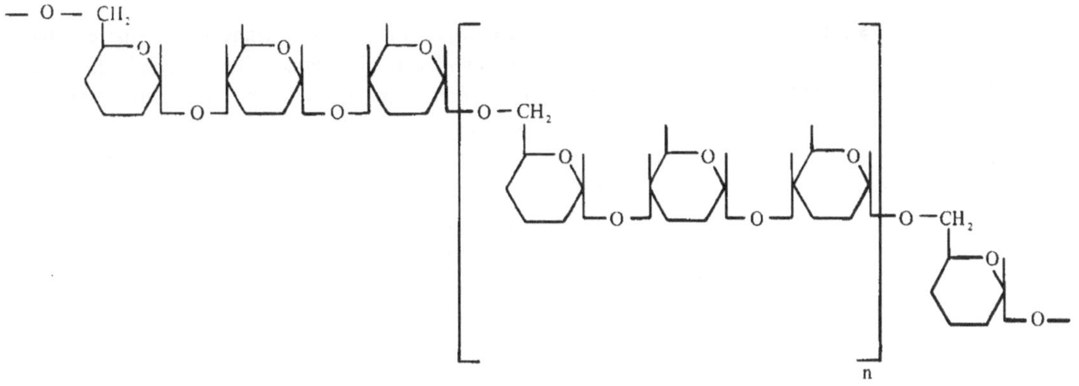

FIGURE 1. Pollulan structure.

cerning ability to utilize cellulose. The joint Food and Agriculture Organization/World Health Organization (FAO/WHO) Expert Committee on Food Additives concluded that cellulose is not digested or absorbed by man.[37] This conclusion was apparently based on the unpublished report of the administration of a single does of $^{14}$C-labeled cellulose to a single volunteer. This result must be weighed against the Olmsted work previously cited and a more recent report of a utilization range of 15 to 87% in 16 volunteer participants in a study using a nonspecific analytical method for measuring cellulose recovery from fecal samples.[38]

Many countries permit use of microcrystalline cellulose powder, a partially depolymerized cellulose in food.[37] Daily intake is unlimited. A slightly cruder form of cellulose, commonly referred to as alpha cellulose, is derived from wood chips and is also commonly used in low calorie food products. The palatability of foods containing appreciable cellulose quantities has always been a problem to the food technologist. In many instances, it is used only at relatively low levels to improve texture. Recent interest in increased intake of dietary fiber has also renewed research on cellulose use in foods. In particular, there have been new attempts to market enriched breads having high fiber content and fewer calories than regular bread. Palatability appears to be satisfactory and preliminary studies in humans indicate weight reduction after regular consumption of the bread.[39]

Pullulan is an extracellular glucan produced by the fungus, *Pullularia pullulan,* that grows on a variety of carbohydrate sources.[40] It is a linear polysaccharide consisting largely of maltotriose units linked by $\alpha$-1 $\rightarrow$ 6 bonds (Figure 1).[41] There have been claims that pullulan functions as a low calorie flour substitute in bread and pastry as well as making transparent biodegradable food wrap.[42] It is also reported to be commercially available in Japan by starch fermentation. Little information is available on its actual utility or metabolism, but a digestibility study implied that it was utilized only about 50% by the rat.[43] Considering the polymer's regular structure and the presence of maltotriose units, low caloric utilization of pullulan appears questionable.

The simplest bulking agents possibly considered for replacement of refined carbohydrates like sugar are the closely related polyols. These compounds have many physical properties similar to monosaccharides and disaccharides. Additionally, they frequently possess a sweet taste approximating that of sucrose. Table 1 summarizes the most pertinent data on polyols and related compounds used or suggested as sugar replacements. As indicated, most of these polyols are not actually low calorie. A discussion of their properties and uses appears warranted because of their historical use in foods. The literature information on caloric utilization of these polyols is confused; the values reported in Table 1 are the best current assessments of this information.

Sorbitol, a normal intermediate in human hexose metabolism, is the most thoroughly investigated of the polyols. It is found in a variety of commonly consumed foods, particularly fruit, and is readily available by glucose hydrogenation. Many regulatory agencies approve its use as a sugar substitute in special dietetic foods. Though it is only half as sweet as sucrose, it has a clean, cool taste attractive in certain foods such as chewing gum and candy. It also finds wide application in

TABLE 1

**Caloric Utilization and Sweetness Characteristics of Polyols**

| Compound | Caloric utilization % | Sweetness (sucrose = 100) |
|---|---|---|
| Sorbitol | 90 | 60 |
| Mannitol | 50 | 50 |
| Xylitol | 100 | 100 |
| Maltitol | 90 | 85 |
| Arabitol | 40 | 50 |
| Isomaltitol | ? | 45 |
| Lactitol | ? | 40 |
| Sorbose | ? | 80 |

toothpaste. As indicated in Table 1, sorbitol has a high caloric value, thus it has no utility in low-caloric foods. Even so, it has been used for many years in foods designed for the diabetic because glucose serum values are not raised appreciably after ingestion of sorbitol. The recent trend towards proper weight maintenance for diabetics plus dietary recommendations of reduced fat but higher carbohydrate intake[44] makes sorbitol use in the diabetic diet less attractive.[45]

Mannitol, found in a few food sources, is commonly obtained as a by-product in the hydrogenation of invert sugar. Like sorbitol, it has regulatory approval for use in special dietetic foods in many countries. Lack of hygroscopicity makes it an excellent bodying agent in powdered foods and a dusting agent for chewing gum. Caloric mannitol utilization is low but variable depending upon dosage.[17] High cost and relatively poor toleration are the principle factors limiting its use. Despite many years of mannitol and sorbitol use in dietetic foods, the FDA (as part of the Generally Recognized as Safe (GRAS) Review decision) now insists that a food which may result in a daily ingestion of 20 g or more mannitol must bear the statement, "Excess consumption may have a laxative effect." Sorbitol, tolerated much better than mannitol, requires a similar label declaration for foods from which the daily sorbitol ingestion is likely to exceed 50 g. Sorbitol and mannitol use in "sugar-free" chewing gum, presumedly less cariogenic than normal chewing gum, has dramatically increased in recent years.

Detailed studies of xylitol made by the University of Turku Dental Research group in Finland culminated in a year-long test involving over 100 subjects. The group's work shows significant reduction in caries formation compared to a sucrose control group when individuals simply consumed five sticks of chewing gum containing xylitol each day.[46] Xylitol, a five-carbon polyol found in a wide variety of fruits and vegetables, is a normal intermediate in the glucuronate-xylulose cycle in man. Consequently, appreciable quantities are normally produced and subsequently metabolized. Since utilization is independent of insulin, xylitol has been used in a number of countries, including the U.S., as a carbohydrate source in special dietetic foods for diabetic individuals and for parenteral nutrition.

Several years ago, concern was expressed over the safety of xylitol when some patients developed adverse reactions after receiving i.v. doses of xylitol solution. Subsequent investigation suggested that the reactions were due to rapid administration of massive doses or to possible solution contamination.[47] This isolated finding of adverse reactions appears to have little relevance to the safety of orally ingested xylitol, especially when judged in light of normal daily production and turnover of appreciable xylitol quantities in the body and the fact that humans have been exposed to high daily dietary intakes of xylitol during the Turku studies without suffering harmful effects. Like all of the polyols and slowly absorbed carbohydrates, xylitol causes gastrointestinal distress if consumed in large doses. In the most definitive study reported, Dubach and co-workers showed that under normal dietary consumption patterns, xylitol was well tolerated even at levels above 100 g/day and that for most volunteers xylitol was tolerated better than sorbitol.[48]

Presently, xylitol has FDA and WHO/FAO clearance as an ingredient in dietetic foods, and the National Institute of Dental Research is preparing to sponsor an elaborate clinical study with children which will compare caries incidence associated with consumption of xylitol and sorbitol chewing gums. Xylitol has excellent sweetness characteristics and formulates easily in foods. The most severe restriction on its use appears to be its relatively high cost. It is usually made by hydrogenation of xylose available from a wide range of xylans. These are found in appreciable quantity in various agricultural waste products; thus, the potential cost of xylitol could become much more attractive to the food processor.

Maltitol (4-α-D-glucopyranosyl-D-sorbitol) is an interesting polyol in that it has excellent sweetness quality approaching that of sucrose, and it is readily available from maltose. The metabolism of maltitol in the mouse[49] and rat[50] implied that the caloric utilization in these species was low. However, maltitol has also produced hyperglycemia in the mouse similar to that caused by an equal amount of sucrose or glucose but at a slower rate.[51] Based on the Japanese studies, maltitol has been used as a bulking agent for low calorie and diabetic foods in Japan. However, work done in Pfizer laboratories shows that it is efficiently utilized by man.[52] The principal utilization pathway was lower gut microbial fermentation.

In a similar fashion, it has been proposed that isomaltitol, available by the hydrogenation of isomaltulose,[53] be used as a low-caloric bulking agent[54] and in diabetic foods.[55] Despite preliminary data in these reports, it appears unlikely that isomaltitol, because of its close structural similarity to maltitol, will be low-caloric in appropriate studies in humans.

Despite its low sweetness level, lactitol (4-β-D-galactopyranosyl-D-glucitol) has been patented as a low-caloric bulking agent.[54] It is available by the hydrogenation of lactose which, in turn, is readily obtained as a by-product in cheese whey separations. No significant metabolism work has been reported on lactitol, but it would be surprising if it had significant low calorie properties. Some evidence suggesting that lactitol could be noncariogenic exists.[57]

Sorbose, though not a polyol, is included in Table 1 because it is the only monosaccharide for which noncariogenic, low-caloric claims have been made.[58] There is little available evidence to support either claim. The compound has a good sweetness level, but since it is usually prepared by sorbitol oxidation, it would be an expensive bulking agent.

Another pentitol, D-arabitol, has been claimed as a low-caloric, nontoxic bulking agent.[59] After oral administration to humans, an average of 63% D-arabitol is excreted in the urine within 24 hr, so the caloric utilization is likely to be less than 40%. Apparently, most of the mammalian dehydrogenase enzymes are unable to attack arabitol, although there is one report of a polyol dehydrogenase from human erythrocytes that can oxidize it to D-ribulose.[60] Arabitol is available from a number of fermentation processes, and it can be produced by the hydrogenation of arabinose, a relatively rare sugar. It apparently is not approved for use as a food additive in any of the major countries.

The remaining water-soluble bulking agents are polysaccharides having no sweet taste. Formulation of these materials into foods in which sugar is substituted creates an additional problem – a nonnutritive sweetener must be used if the food is to retain its customary sweetness level. The regulatory controversy over nonnutritive sweeteners has been going on since 1968 and will probably continue for some time to come. It makes food-applications work with these substances exceedingly frustrating for the industrial food scientist.

Arabinogalactan (Larch Gum), a highly-branched polymer of arabinose and galactose obtained from the Western Larch tree, has a caloric utilization value of 1.3 cal/g in a digestibility test in rats.[61] Though it has FDA clearance for use in foods and has many suitable physical properties such as good water solubility with the solutions having low viscosities, actual use has been small due to taste, functional characteristics, and heat-stability problems.

A variety of other plant and microbial polysaccharides have received some attention as low calorie bulking agents. Information on their digestibility is summarized in an excellent review by Jeanes.[62]

Finally, mention should be made of polydextrose, a Pfizer research product.[63] It is a bland tasting, water-soluble polymer with low-caloric utilization. Many of its physical properties such as solubility, viscosity, and melt characteristics resemble those of sucrose. In a wide range of applications, it can provide the bulk and texture normally obtained from sucrose, and in some applications, it can function as a partial replacement for fat, flour, or starch. Some of the more interesting examples of polydextrose utility include: low calorie, sugar-free hard candy, frozen dessert products containing only 3% butterfat with an ice cream-like texture; and a fat-free salad dressing.[64]

Polydextrose is a randomly bonded polymer of glucose containing some sorbitol and citric acid. It is very resistant to digestive enzymes.[65] Studies in humans, using radiolabeled material, have shown that it is only about 25% utilized, so it has a caloric value of only 1 cal/g.[66] Therefore, in the foods mentioned above and in many other applications, the caloric reduction of the polydextrose-based foods can exceed 50% with retention of good organoleptic properties.

An insoluble version of polydextrose possibly functioning as a flour or starch replacement has also been prepared.

## SUMMARY

Obesity is the most important nutritional problem in the U.S. today. The need for low calorie foods and the utility of bulking agents in their manufacture is clear. However, technical limitations of the presently approved list of bulking agents restrict their use to relatively few foods. Some of the newer bulking agents not yet commonly used in foods may fill these needs. Still, a large research and development effort is needed before the overweight consumer will have a variety of highly acceptable foods available at a cost that is suitable to a normal food budget.

**Acknowledgments**

The author would like to express his appreciation to his colleagues, Drs. C. R. Stephens, H. H. Rennhard, R. P. Allingham, G. D. Gruetzmacher and to Mr. A. Torres for constructive suggestions and criticisms during the compilation of this chapter.

## REFERENCES

1. **Bushkirk, E. R.,** Obesity: A brief overview with emphasis on exercises, *Fed. Proc.,* 33, 1948, 1974.
2. **Mayer, J.,** Obesity, *Postgrad. Med.,* 55, 66, 1972.
3. **Davidson, J. K.,** Controlling diabetes mellitus with diet therapy, *Postgrad. Med.,* 59, 144, 1976.
4. **Mayer, J..,** *Overweight: Causes, Cost and Control,* Prentice Hall, Englewood Cliffs, N.J., 1968, 2.
5. **Flatt, J. P. and Blackburn, G. L.,** The metabolic fuel regulatory system: Implications for protein-sparing therapies during caloric deprivation and disease, *Am. J. Clin. Nutr.,* 27, 175, 1974.
6. **Van Itallie, T. B. and Campbell, R. G.,** Multidisciplinary approach to the problem of obesity, *J. Am. Diet. Assoc.,* 61, 385, 1972.
7. **Paulsen, B. K., Lutz, R. N., McReynolds, W. T., and Kohrs, M. B.,** Behavior therapy for weight control: Long-term results of two programs with nutritionists as therapists, *Am. J. Clin. Nutr.,* 29, 880, 1976.
8. **Danowski, T. S.,** The management of obesity, *Hosp. Prac.,* p. 39, April 1976.
9. Diet and Coronary Heart Disease, American Heart Association, New York, 1973.
10. **Kuo, P. T.,** Hyperlipidemia and coronary artery disease, *Med. Clin. North Am.,* 58, 351, 1974.
11. **Lutwak, L.,** The role of dietary calcium: phosphorus ratio in human nutrition, *Proc. Meat Ind. Res. Conf.,* p. 63, 1975.
12. **Klose, R. E. and Glicksman, J.,** Gums in *Handbook of Food Additives,* 2nd ed., Furia, T. E., Ed., The Chemical Rubber Co. Press, 1972, 295.
13. **Merrill, A. L. and Watt, B. K.,** Energy value of foods – Basis and derivation, Agricultural Handbook No. 74, U.S. Department of Agriculture, 1955.
14. **Rice, E. E., Warner, W. D., Mone, P. E., and Poling, C. E.,** Comparison of the metabolic energy contributions of foods by growth under conditions of energy restriction, *J. Nutr.,* 61, 253, 1957.
15. **El-Harith, E. A., Dickerson, J. W. T., and Walker, R.,** Potato starch and caecal hypertrophy in the rat, *Food Cosmet. Toxicol.,* 14, 115, 1976.
16. **Hefebvre, P., Mosora, F., Lacroix, M., Luyckx, A., Lopez-Habib, G., and Duchesne, J.,** Naturally labeled $^{13}$C-glucose. Metabolic studies in human diabetes and obesity, *Diabetics,* 24, 185, 1975.
17. **Nasrallah, S. M. and Iber, F. L.,** Mannitol absorption and metabolism in man, *Am. J. Med. Sci.,* 258, 80, 1969.
18. **Stephenson, L. S. and Latham, M. C.,** Lactose intolerance and milk consumption: The relation of tolerance to symptoms, *Am. J. Clin. Nutr.,* 27, 296, 1974.
19. **Madsen, J.,** Ingredient requirements, development techniques for low-calorie table spreads, *Food Prod. Dev.,* 10, 72, 1976.
20. **Babayan, V. K.,** Modification of food to control fat intake, *J. Am. Oil Chem. Soc.,* 51, 260, 1974.
21. **Pyke, M.,** Synthetic and artificial food, *Nutr. Food Sci.,* 28, 18, 1972.
22. **Rizzi, G. P. and Taylor, H. M.,** Synthesis of higher polyol fatty acid polyesters, U.S. Patent 3,963,699, June 15, 1976.

23. **Minich, A.,** Dietetic compositions, U.S. Patent 2,962,419, November 29, 1960.
24. **Mattson, F. H. and Volpenhein, R. A.,** Hydrolysis of fully esterified alcohols containing from one to eight hydroxyl groups by the lipolytic enzymes of rat pancreatic juice, *J. Lipid Res.*, 13, 325, 1972.
25. **Mattson, F. H. and Nolen, G. A.,** Absorbability by rats of compounds containing from one to eight ester groups, *J. Nutr.*, 102, 1171, 1972.
26. **Mattson, F. H., Jandacek, R. J., and Webb, M. R.,** The effect of a non-absorbable lipid, sucrose polyester, on the absorption of dietary cholesterol by the rat., *J. Nutr.*, 106, 747, 1976.
27. **Fallat, R. W., Glueck, C. J., Mattson, F., and Lutmer, R.,** Sucrose Polyester: A cholesterol lowering, noncaloric, unabsorbable, synthetic, fat food additive, *Clin. Res.*, 23, 319A, 1975.
28. **Merten, H. L.,** Low calorie lipids, *J. Agric. Food Chem.*, 18, 1002, 1970.
29. **Gros, A. T. and Feuge, R. O.,** Properties of the fatty acid esters of amylose, *J. Am. Oil Chem. Soc.*, 39, 19, 1962.
30. **Booth, A. N. and Gros, A. T.,** Caloric availability and digestibility of new-type fats, *J. Am. Oil Chem. Soc.*, 40, 551, 1963.
31. **Sucher, R. W., Robbins, E. A., Sidoth, D. R., Schuldt, E. H., and Seeley, R. D.,** U.S. Patent 3,867,554, 1975.
32. **Spiller, G. A. and Amen, R. J.,** Dietary fiber in human nutrition, *Crit. Rev. Food Sci. Nutr.*, p. 39, November 1975.
33. **Williams, R. D. and Olmsted, W. H.,** The manner in which food controls the bulk of the feces, *Ann. Intern. Med.*, 10, 717, 1936.
34. **Reinhold, J. G., Faradji, B., Abadi, P., and Ismail-Beigi, F.,** Decreased absorption of calcium, magnesium, zinc and phosphorus by humans due to increased fiber and phosphorus consumption as wheat bread, *J. Nutr.*, 106, 493, 1976.
35. **Conrad, H. E., Watts, W. R., Iacono, J. M., Kraybill, H. F., and Friedmann, T. E.,** Digestibility of uniformly labeled carbon-14 soybean cellulose in the rat, *Science*, 127, 1293, 1958.
36. **Johnson, R. B., Peterson, D. A., and Tolbert, B. M.,** Cellulose metabolism in the rat, *Nutrition*, 72, 353, 1960.
37. Toxicological evaluation of some enzymes, modified starches and certain other substances, World Health Organization (WHO) Food Additive Series, Geneva, Switzerland, No. 1, 85, 1972.
38. **Miltov Thompson, F. J. and Lewis, B.,** The breakdown of dietary cellulose in man, *Gut*, 12, 853, 1971.
39. **Anon.,** Let them eat bread! Watch them lose weight!, *Med. World News*, p. 36, December 15, 1975.
40. **Bender H., Lehmann, J., and Wallenfels, K.,** Pullulan, Ein extracelluläres glucan von pullaria pullulans, *Biochim. Biophys. Acta*, 36, 309, 1959.
41. **Catley, B. J. and Whelan, W. J.,** Observations on the structure of Pullulan, *Arch. Biochem. Biophys.*, 143, 138, 1971.
42. **Anon.,** Japanese develop starch-derived plastic, *Chem. Eng. News*, 51, 40, 1973.
43. **Kato, K. and Shiosaka, M.,** Food compositions containing pullulan, U.S. Patent, 3,875,308, April 1, 1975.
44. **Kiehm, T. G., Anderson, J. W., and Ward, K.,** Beneficial effects of a high carbohydrate, high fiber diet on hyperglycemic diabetic men, *Am. J. Clin. Nutr.*, 29, 895, 1976.
45. **Lenner, R. A.,** Specially designed sweeteners and food for diabetics – A real need?, *Am. J. Clin. Nutr.*, 29, 726, 1976.
46. **Scheinin, A., Mäkinen, K. K., Tammisalo, E., and Rekola, M.,** Turku sugar studies. XVIII. Incidence of dental caries in relation to 1-year consumption of xylitol chewing gum, *Acta. Odontol. Scand.*, 33, 269, 1975.
47. **Thomas, D. W., Edwards, J. B., and Edwards, R. G.,** Toxicity of parenteral xylitol, in *Sugars In Nutrition*, Sipple, H. L. and McNutt, K. W., Eds., Academic Press, New York, 1974, 567.
48. **Dubach, V. C., Finer, E., and Forgo, I.,** Orale vërtraglichkeit von Xylit bei stoffwechselgesunden Probanden, *Schweiz. Med. Wochenschr.*, 99, 190, 1969.
49. **Kamoi, M., Shimizu, Y., Kawauchi, M., Fujii, Y., Kikuchi, T., Mizukawa, S., Yoshioka, H., Kibata, M., and Mitsuhashi, M.,** Maltitol metabolism, *Igaku No Ayumi*, 82, 208, 1972; *C. A.*, 78, 55816, 1973.
50. **Oku, T., Inoue, Y., and Hosoya, N.,** Absorption and excretion of maltitol-U-$^{14}$C in rat, *Eiyo To Shokuryo*, 24, 399, 1971; *C. A.*, 76, 95260, 1972.
51. **Lederer, J., Delville, P., and Crevecoeur, E.,** New sweeting agent. Maltitol, *Sucr. Belge Sugar Ind. Abstr.*, 93, 311, 1974; *C. A.*, 81, 167926, 1975.
52. **Rennhard, H. H. and Bianchine, J. R.,** Metabolism and caloric utilization of orally administered maltitol-$^{14}$C in rat, dog and man, *J. Agric. Food Chem.*, 24, 287, 1976.
53. **Schieweck, H., Steinle, G., and Haberl, L.,** Low caloric sweetener and sweetener base, U.S. Patent 3,865,957, February 11, 1975.
54. **Musch, K., Siebert, G., Schieweck, H., and Steinle, G.,** Nutritional and physiological investigations of isomaltitol in rats, *Ernaerungswiss. Suppl.*, 15, 3, 1973.
55. **Siebert, G., Grupp, U., and Heinkel, K.,** Studies on isomaltitol, *Nutr. Metab.* 18 (Suppl. 1), 191, 1975.
56. **Hayashibara, K.,** Foodstuffs sweetened with lactitol, British Patent 1,253,300, November 10, 1971.
57. **Malsuo, T.,** Effects of maltitol and lactitol on lactic acid production by oral microorganisms, *Shigaku*, 60, 760, 1973; *C. A.*, 81, 58745, 1974.
58. **Huchette, M. and Leroy, P.,** Sugar substitute, German Patent 2,308,163, 1974; *C. A.*, 81, 76699, 1974.
59. **Crick, R. G. D.,** Improvements in or relating to sweetening agents for food, British Patent 884,961, December 20, 1961.

60. **Wang, Y. M. and van Eys, J.,** The enzymatic defect in essential pentosuria, *N. Engl. J. Med.*, 282, 892, 1970.
61. **Anon.,** *Chem. Eng. News,* 32, 37, 1965.
62. **Jeanes, A.,** Digestibility of food polysaccharides by man, in *Physiological Effects of Food Carbohydrates,* American Chemical Society Symposium Series No. 15, Jeanes, A. and Hodge, J., Eds., American Chemical Society, Washington, 1975.
63. **Rennhard, H. H.,** Polysaccharides and their preparation, U.S. Patent 3,766,165, October 16, 1973.
64. **Torres, A.,** personal communication.
65. **Rennhard, H. H.,** personal communication.
66. **Figdor, S. K. and Rennhard, H. H.,** personal communication.

Chapter 4

# THE PRACTICAL REQUIREMENTS FOR THE USE OF SYNTHETIC SWEETENERS

K. M. Beck

## TABLE OF CONTENTS



## I. INTRODUCTION

Many different things taste sweet. If one pauses to think about well-known sweeteners – sugars like sucrose, glucose, and fructose – polyols like sorbitol and mannitol – amino acids like glycine – synthetic chemicals like cyclamate and saccharin – it becomes apparent that a great variety of sweet natural and synthetic chemicals exists. This variety of chemical structures offers a fascinating spectrum of technical properties, but there are a number of practical considerations greatly limiting the actual use of these chemicals as sweetening agents. There are frequent reports of new synthetic chemicals having a sweet taste. While considering these new synthetic sweeteners as potential food additives, it is important to realize their practical requirements for use.

Sweetness is one of four basic taste sensations: sweetness, sourness, bitterness, and saltiness. Sourness and saltiness are fairly well correlated with specific chemical structures. A sour taste is associated with hydrogen ions at the receptor site; thus, it is a common taste for protonic acids, such as citric, malic, acetic, tartaric, or phosphoric acid. The saline taste is characteristic of halides, nitrates, and sulfates of sodium, potassium and lithium. It is limited to soluble, ionizable salts. Chemically, one doesn't stray very far from sodium chloride and still obtain a true salty taste.

Bitterness and sweetness are not confined to such well-defined chemical classes. A variety of chemicals offers bitter taste. Alkaloids, such as quinine, and xanthines, like caffeine, are bitter. These are heterocyclic nitrogen compounds. However, bioflavenoids, such as naringin, lack nitrogen and taste bitter. Antibiotics like penicillin and erythromycin are bitter. Aspirin is bitter, and antihistaminics have a persistent bitterness that is hard to mask.

The structural variety or nonspecificity of chemicals with sweet taste is as great as for chemicals with bitter taste. The popular use of sucrose, glucose, lactose, maltose, sorbitol, and mannitol as sweeteners promotes a tendency to think of carbohydrates and polyols as the main class of sweet-tasting chemicals. Noting that glycerin has a sweetness comparable to that of glucose, one might think that sweetness is a characteristic of molecules having multiple HCOH groups (Table 1). However, of all the monosaccharides from trioses to hexoses and the multitude of known disaccharides and polysaccharides, only a few are sweet. Most are tasteless. There are even bitter tasting sugars. Gentiobiose is bitter. Mannose provides an interesting example of taste stereospecificity. α-D-mannose is sweet, and the stereoisomer β-D-mannose is bitter.

Let's look more closely at sweet-tasting chemicals that are not carbohydrates. There are several inorganic salts with sweet taste, especially the formate, acetate, and propionate salts of lead and beryllium. Lead acetate, known for centuries as "sugar of lead," is perhaps the most famous. It is quite toxic and once had nefarious uses as a sweetening agent in preparing fatal cocktails. Several amino acids are sweet: glycine, histidine, leucine, tyrosine, phenylalanine, and tryptophan. Stereoisomers of the same compound often taste different, and sweetness particularly seems to be stereospecific. With all of these amino acids, the D-isomers are sweet and the L-isomers are bitter.[1] The most striking example is tryptophan; the D-isomer is 35 times as sweet as sucrose, and the L-isomer is not sweet at all. Several amino acid derivatives have pronounced sweetness. Aspartame is a methyl ester of the dipeptide, aspartylphenylalanine and is approximately 180 times as sweet as sugar. The 6-chloroderivative of D-tryptophan is reported to be 1000 to 1300 times as sweet as sugar. A number of natural products have pronounced sweetness: glycyrrhizin from licorice root; stevioside from an herb that grows in Paraguay; protein fractions called Miraculin, Monellin, and Thaumatin isolated respectively from Miracle fruit (*Synsepalum dulcificum*), Serendipity berries (*Dioscoreophyllum cumminsii*), and Katemfe (*Thaumatoccus daniellii*).[13,16]

A host of completely synthetic organic chemicals are known to taste sweet. Reports of more than 200 such compounds reveal two interesting characteristics. First is the great disparity of structures; they do not belong to a single chemical class or have any common nucleus or functional group. Historically, chloroform was the first synthetic chemical reported to be sweet (in 1831). Dulcin, *p*-ethoxyphenylurea, was discovered in 1883, and saccharin in 1879. Consequently, the field of synthetic sweeteners is about 100 years old. Since that time, sweetness has been reported for such varied structures as alkoxynitroanilines, benzimidazoles, triazoles and tetrazoles, alkenols, hydantoins, dihydrochalcones, and sulfamates. It is interesting that most were discovered accidentally. Theories to explain why these chemicals taste sweet have failed in predictive value.[20] Sweetness intensity is the other startling characteristic of synthetic compounds. They often are several hundred times as sweet as sucrose (Table 2). Carbohydrate sweeteners and the structurally related polyols vary in sweetness over a narrow range, from about one third sucrose sweetness to about 20% greater sweetness than sucrose (Table 1). The synthetic sweeteners range from about 15 to 4000 or more times the sweetness of sucrose (Table 2).

TABLE 1

**Relative Sweetness of Carbohydrate Sweeteners**

| Sweeteners | Sweetness relative to sucrose as 1.0 |
|---|---|
| Sucrose (standard) | 1.0 |
| Glucose | 0.7 |
| Fructose | 1.1 |
| Lactose | 0.4 |
| Maltose | 0.5 |
| Sorbitol | 0.5 |
| Mannitol | 0.7 |
| Glycerin | 0.7 |
| Cellobiose | 0.0 |
| α-D-Mannose | Sweet |
| β-D-Mannose | Bitter |
| Isomaltose | Sweet |
| Gentiobiose | Bitter |

Obviously, there is no shortage of chemicals suitable for consideration as synthetic sweeteners. Let's look at the practical criteria limiting this list to a few useful products. Of course, these arise from where and how the materials can be used.

Sugar is a carbohydrate, a nutrient, and is

generally useful as a source of sweetness in foods and drugs. A synthetic sweetener can be used in foods and drugs thereby replacing the taste provided by sugar and offering some dietary or technological advantages over use of sugar. A comparison of cyclamate and sucrose properties (Table 3) illustrates some differences between sugars and synthetic sweeteners. These characteristic differences can provide bases for practical uses of synthetic sweeteners.[7] There are two dietary reasons for using a nonnutritive synthetic sweetener in foods and beverages (Table 4): to provide sweetness without an insulin requirement in foods for diabetics and to provide noncaloric sweetening in formulation of low calorie foods for people on calorie-restricted diets.

TABLE 2

**Relative Sweetness of Noncarbohydrate Sweet Chemicals**

| | |
|---|---|
| 3-Methylcyclopentylsulfamate | 15 |
| *p*-Anisylurea | 18 |
| Sodium cyclohexylsulfamate (cyclamate) | 30 |
| Chloroform | 40 |
| Methoxy-2-amino-4-nitrobenzene | 167 |
| *p*-Ethoxyphenylurea (Dulcin) | 200 |
| 6-Chlorosaccharin | 200 |
| Sodium benzosulfimide (saccharin) | 300 |
| *n*-Hexylchloromalonamide | 300 |
| Stevioside | 300 |
| Naringin dihydrochalcone | 300 |
| 2-Amino-4-nitrotoluene | 300 |
| *p*-Nitrosuccinanilide | 350 |
| *p*-Methoxymethylnitrobenzene | 500 |
| 1-Bromo-5-nitroaniline | 700 |
| 6-Chloro-D-tryptophane | 1000 |
| Perillaldehyde oxime (perillartine) | 2000 |
| 5-Nitro-2-n-propoxyaniline (P-4000) | 4000 |
| Neohesperidine dihydrochalcone | 6000 |

*Note:* All entries are compared to a standard sweetness of sucrose = 1.

In a number of technological applications, the properties of synthetic sweeteners provide advantages over the practical limitations of sugar (Table 4). In pharmaceuticals, synthetic sweeteners frequently are used as flavoring agents to mask the taste of drugs and to improve palatability of syrups, elixirs, chewable tablets, and tablet coatings.[12,17] For this purpose, cyclamate especially is qualitatively superior to sugars. Cyclamate and saccharin are useful flavoring agents in toothpaste and mouthwash. Cyclamate is also an excellent flavoring agent for grapefruit juice.

Compared to synthetic sweeteners, sugar has a low order of sweetness and is a bulky sweetener. The amount of sugar used to sweeten fruit sections produces a higher osmotic pressure in the packing syrup than in the fruit. The flow of liquid achieving equilibration causes fruit shrinkage. The low concentration required in using a synthetic sweetener reverses this pressure balance, provides a plumping effect in the fruit, and prevents flavor and texture loss.[19] An intense sweetening agent can be used to prepare compact foods. In pharmaceuticals, it can be used to make smaller chewable tablets and to flavor resin coatings for tablets.[12] Some sugars are hygroscopic and most of them are difficult to crystallize and dry. This makes freeze-drying difficult in fruit processing. In pharmaceuticals it causes cap-lock problems.[17] A synthetic sweetener that is easy to dry and nonhygroscopic can obviate these problems. Sugar does not have a very high decomposition point. It tends to caramelize and char during high-temperature cooking processes like frying. A synthetic sweetener with greater heat stability can be used to prepare bacon that will brown without charring when fried.[4] Sugar serves as food for most microorganisms. Synthetic sweeteners that are microbiologically inert can be used to make noncariogenic chewing

TABLE 3

**Comparison of Sucrose and Cyclamate Properties**

| Properties | Sucrose | Sodium cyclamate |
|---|---|---|
| Formula | $C_{12}H_{22}O_{11}$ | $C_6H_{11}NHSO_3Na$ |
| Chemical class | Carbohydrate | Salt |
| Sweetness | 1 | 30–60 |
| Melting point | 160–180° (decomposes) | 480–500° (decomposes) |
| Solubility in water | 68 g/100 cm² | 21 g/100 cm² |
| Specific gravity | 1.176 (40% solution) | 1.002 (1% solution) |
| Viscosity | 5.187 (40% solution) | 1.04 (1% solution) |
| Food value | 4 cal/g | 0 |

gum or to flavor fermented products like sweetened yogurt. Considering these uses, what are the characteristics required of a synthetic sweetening agent?

### A. Sensory Properties

Sugar used in foods provides a sweet taste. People are accustomed to a certain sweetness quality of a certain time-intensity curve for this taste. To insure good acceptability in foods and beverages, a sugar substitute should have a sweetness profile reasonably similar to that of sugar and should elicit only a sweetness perception, that is, be fairly free of bitterness or other tastes. Ideally, it should have a sweet taste apparent in a second or two and remaining for about 30 sec (Figure 1).

Chemicals with off-tastes or aftertastes present complicated flavor problems. If the sweetness response is delayed or if it persists for long periods, it will cause a product to taste "funny." The more a sweetener tastes like sugar, the easier it is to use in formulation of food and beverage products having broad consumer acceptance. Many synthetic chemicals with sweet taste also have other tastes or unusual sweetness profiles, and are poor candidates for food uses.

TABLE 4

**Uses for Synthetic Sweeteners**

| | |
|---|---|
| **Dietary** | |
| Calorie reduction – provide sweetness without calories in formulating low calorie foods and beverages for people on calorie-controlled diets | |
| Sugar replacement – provide sweetness without insulin requirement in products for diabetics | |
| **Technological** | |
| Flavoring agent | |
| High intensity sweetener | |
| Nonhygroscopic sweetener | |
| Noncharring sweetener | |
| Microbiological resistance | |

### B. Chemical Properties

Contrary to appearances created by the way some feeding studies are conducted, people don't sit down and eat bowls of sugar or spoonfuls of synthetic sweetener. Usually sweetness is only one component of a complex flavor system and is used along with a variety of natural and artificial flavors and colors. It must be compatible with these various chemical compounds. For adequate shelf stability, a sweetening agent should be chemically

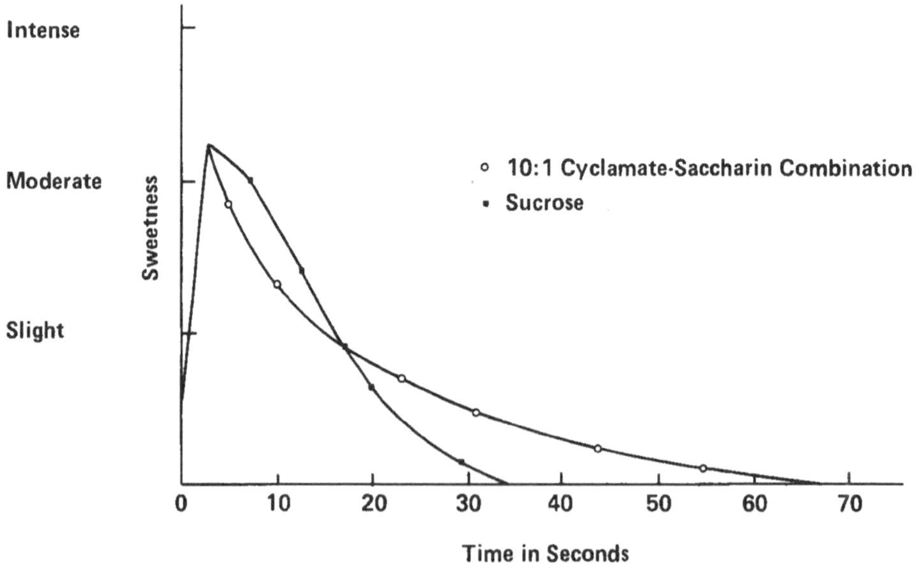


FIGURE 1. Both sucrose and cyclamate-saccharin in combination have above moderate sweet taste at 2 to 3 sec. Sucrose has a rounded sweetness profile disappearing in about 35 sec. Sweetness of the cyclamate-saccharin combination drops off somewhat faster, but total taste lasts for about 60 sec. (From McCormick, R. D., Ed., Food Prod. Dev., May, 1975, Copyright, 1975, Food Product Development. With permission.)

unreactive or inert toward all natural constituents and chemical additives in products in which it is used. In pharmaceuticals, a sweetener must be compatible with a variety of active drugs.

### C. Physical Properties

Temperature stability is an important requirement for chemical sweeteners. Food processing often involves cooking; a sweetener should be stable toward boiling, baking, and pressure cooking. Likewise, it must be stable toward low temperature processing such as fast freezing and freeze-drying.

Foods must be stored, shipped, and later stocked on grocery shelves. A sweetener should be stable under various actual storage conditions. Length of stability will vary with the food product, but a 2-year shelf life is necessary for some, and 6 months is minimal for most processed food products.

Many foods are liquid, have a high moisture content, or go through a liquid phase in processing. The stability of a sweetener in solution is important. The pH of foods and beverages can range from 2.5 to 8, so stability throughout this pH range is essential for broad application.

Though many foods contain sugar concentrations of 15% or less, products like jellies and syrups may contain more than 50% sugar. To be widely applicable, a synthetic sweetener should be sufficiently soluble to duplicate the sweetness of simple syrup which contains about 70% sugar. Sugars are carbohydrates; therefore, they have certain chemical and physical properties characteristic of these structures. As discussed earlier, there are technological uses for which a synthetic sweetener that is not a carbohydrate may have properties superior to those of sugar.

### D. Safety

Proof of safety required for any new food additive presents a difficult problem in use of sweetening agents. It is necessary to feed experimental animals sufficiently high levels of sweetener to produce some adverse effect. Then studies are conducted to determine the maximum no-effect level. Requirement of a 100-fold safety factor over this maximum no-effect level makes it difficult to qualify practical amounts of a food additive. The Zero Tolerance Concept of the Delaney Clause,* particularly, is of concern to scientists. Studies in metabolism and excretion must be conducted, and separate feeding studies done with any metabolites.

Food technologists are familiar with the great number of natural toxicants in food.[10] It is exasperating for scientists to see consumer protectionists agitate for safety requirements for food additives that are, in effect, less toxic than many chemicals that occur naturally in foods. This propagates a double standard – one for natural substances and one for synthetic chemicals. A synthetic sweetening agent elicits especially stringent safety requirements that sucrose and lactose certainly never could meet. Consider that the potato is one of the most widely eaten foods; salt is the most widely used chemical in food; and aspirin is the most widely used drug. Yet with current requirements for proof of safety, one might experience serious difficulty in obtaining Food and Drug Administration (FDA) clearance for the potato as a new food, for sodium chloride as a new food additive, or for acetylsalicylic acid as a new drug. The risk/benefit concept seems to need some development. Presently, for even a fairly innocuous chemical, one should plan on 10 years and more than $10 million to collect the data necessary for a food additive petition for a new sweetener – then prepare to battle continuously with self-appointed "consumer protectionists" who seem to resist anything new. Applying these practical considerations of safety, chemical, physical, and sensory properties to the list of known, sweet-tasting chemicals will indicate which are presently or potentially useful.

## II. CHEMICAL SWEETENERS

### A. Saccharin

This is the only synthetic sweetener now approved for wide range food additive use. In 1879, saccharin (Structure 1) was discovered by Remsen and Fahlberg at Johns Hopkins University and has been in use continuously since 1900. It is

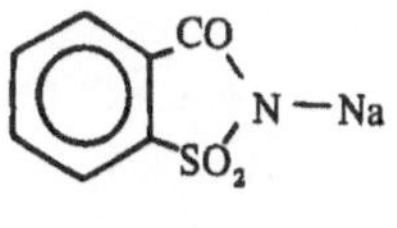


Structure 1

*Food Additive Amendment of the Federal Food, Drug and Cosmetic Act, 1958.

300 to 400 times as sweet as sucrose. Chemical and physical properties are quite satisfactory; it has good compatibility, solubility, and stability. For many people, saccharin has a bitter aftertaste which has always been a limitation to its acceptability. Saccharin is a good component for a co-sweetener combination in which another synthetic sweetener masks the saccharin aftertaste. However, aftertaste is a limiting factor for use of saccharin alone.

**B. Cyclamate**

This sweetener was discovered at the University of Illinois in 1937 and used commercially in the U.S. from 1950 until further use was banned in 1969. Cyclamate use is still allowed in more than 30 countries around the world. It has excellent physical and chemical properties and good taste characteristics. Cyclamate (Structure 2) is 30 to 60 times as sweet as sucrose. Beginning about 1955 there was increasing use of cyclamate-saccharin combinations, particularly in a mixture in which each contributed about half of the sweetness. This permitted food technologists to formulate highly acceptable, low calorie and sugar-free products.[2] A food additive petition was filed in November 1973 permitting cyclamate use in the U.S. again. This petition was denied on October 4, 1976. Abbott Laboratories, North Chicago, Illinois, has requested that an administrative law judge be appointed as a hearing examiner in this matter.

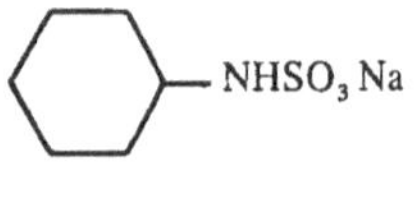


Structure 2

**C. Aspartame**

Aspartyl-phenylalanine methyl ester (Structure 3) developed by G. D. Searle & Co., Skokie, Illinois, was approved in 1974 for use in dry products. Approval was stayed in November, 1975, and the safety data are still being reviewed. Although not completely noncaloric, it has a

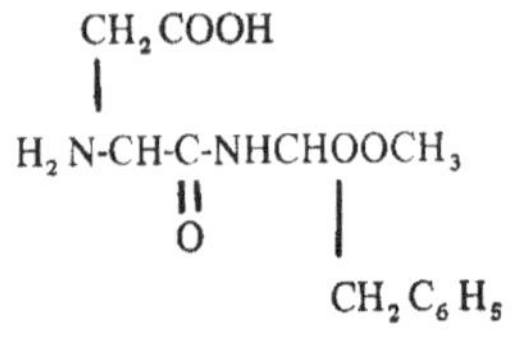


Structure 3

sweetness of 140 to 180 times that of sugar, so its calorie contribution is negligible. Aspartame hydrolyzes in water; therefore, physical stability in liquid products is a serious practical limitation. In neonatal mice, the question of brain damage due to aspartic acid from aspartame acting with glutamic acid illustrates the complexity of questions that can be raised regarding the safety of sweetening agents.

**D. Glycyrrhizin**

Glycyrrhizin is not a synthetic chemical, but a nonnutritive sweetener. Glycyrrhizin, an extract of licorice root, is 50 to 100 times as sweet as sugar. Its use is permitted in food, but the pronounced licorice taste limits its use to specific products in which this taste can be tolerated.

**E. Dihydrochalcones**

About 15 years ago, Horowitz and Gentili at the United States Department of Agriculture (USDA) in Pasadena, California, developed a series of sweet-tasting dihydrochalcones that can be made from citrus bioflavanoids.[15] Two of these were interesting enough to stimulate further work – naringin dihydrochalcone and neohesperidine dihydrochalcone (Structure 4). Solubility would be a limiting factor for the naringin derivative. Taste is a limiting factor for both. The dihydrochalcones have delayed taste, and sweetness persists for several minutes. They also have a characteristic aftertaste. Toxicity work is still being done by the USDA. However, if they are cleared for use, the dihydrochalcones will probably find use only in certain confections and in pharmaceuticals in which the taste characteristics are acceptable.

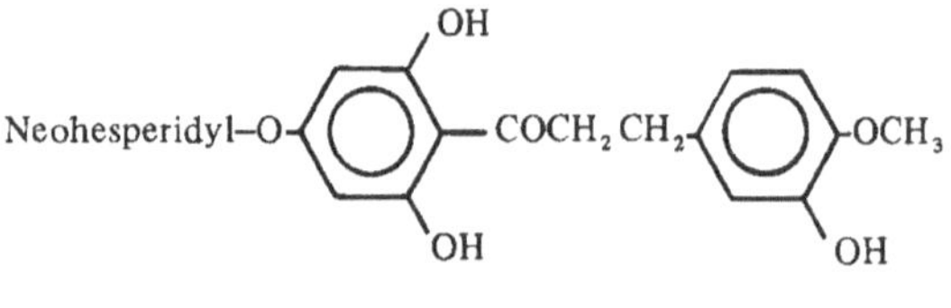


Structure 4

## III. OTHER SWEETENING AGENTS

Several sweetening agents have been studied extensively,[7,13,21] and a few even used briefly in the past. However, their toxicity has been discovered. These include Dulcin, P-4000, perillaldehyde oxime, and stevioside. None of these sweeteners will likely be approved for use in the U.S.

because they do not meet safety requirements. The sweet proteins, Monellin and Thaumatin, have serious stability limitations. Miraculin is not actually a sweet-tasting substance but a taste-modifying chemical eliciting a sweetness response from tart foods. Use of this protein would require educating consumers to accept entirely new flavor characteristics. Therefore, stability and taste limitations exist for use of these proteins as sweeteners.

Some new synthetic sweeteners are being studied currently. One that has attracted attention is Acetosulfam (Structure 5).[9] Discovered at Farbwerke Hoechst A. G. (Frankfurt am Main, Germany), it is reported to have very low acute toxicity. Preliminary reports (not yet published) state that it has a taste similar to that of saccharin and is about half as sweet. Chronic toxicity work is underway, but Acetosulfam is still at an early stage of development.

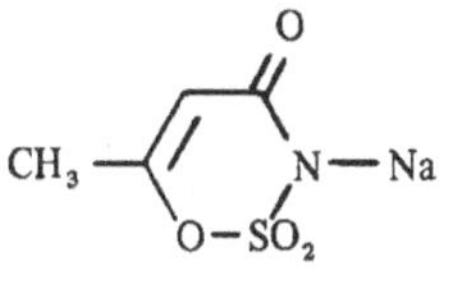


Structure 5

The perfect synthetic sweetening agent is yet to be found, but from a technical viewpoint, a cyclamate-saccharin combination makes a very good, practical sweetener. Its taste characteristics are close enough to sugar to allow formulation of products with excellent consumer acceptance. Chemical and physical stability are very good. The safety of cyclamate and saccharin has been studied extensively, and scientists generally agree that both sweeteners are harmless under actual conditions of use. Furthermore, these sweeteners are economically feasible. Technically, they are the sweeteners to beat.

There are advantages to having several synthetic sweeteners available so that one can select the best agent for each specific application. If saccharin, cyclamate, aspartame, and neohesperidine dihydrochalcone all were permitted, one would have a wide spectrum of useful sweeteners for new product development.

## IV. THE NEED FOR SYNTHETIC SWEETENERS

If sucrose, glucose, and high fructose syrups are available, then why is there need for synthetic sweeteners? Certainly, there are three existing needs, and a fourth will become important in the future.

First, there are over 10 million diabetics in the U.S. A generation ago, a diabetic was commonly told to forego sweetness. A good synthetic sweetener allows diabetics palatable diets with little or no insulin requirement. Therefore, synthetic sweeteners are useful agents in total management of diabetes.

Second, obesity has been repeatedly cited as the nation's leading health problem. Adequate statistical evidence shows that longevity is inversely proportional to overweight. The Food and Nutrition Board of the National Academy of Sciences recommended in 1968 that all adults in the U.S. reduce their caloric intake.[22] Recommended calorie levels are based on age and height, but the average middle-aged male, for example, should reduce intake from 3000 to 2600 cal and the comparable female from 2200 to 1850 cal/day. Deleting 100 g of sugar per day from the diet would be a 400-cal reduction – the amount recommended by the Food and Nutrition Board. If synthetic sweeteners permit some of this reduction without sacrificing palatability, their use should be an allowable option. For many persons, watching caloric intake is a way of life. It should be permissible for them to enjoy noncaloric soft drinks and low calorie desserts with noncaloric sweeteners if they so desire.

Third, as mentioned above, sugar does have limitations due to its physical properties. Synthetic sweeteners have technical advantages in certain products, especially pharmaceuticals, but also in certain foods like bacon and canned fruit. Technical superiority should be recognized and allowed.

The fourth need for synthetic sweeteners apparently contradicts the second, but there are contradictions in this world. Food is readily available in the U.S., and overeating is a problem here. However, half the world's people do not have enough to eat, and death by starvation is a fact in many poor countries. Earth is a planet of fixed dimensions with a finite amount of arable land. As the total world population expands, man must find new and better ways to provide food. Food technologists certainly are familiar with efforts to develop protein sources – single-cell protein from methyl alcohol or petroleum, fish flour, and edible algae. Flavoring these materials will be important, and sweetness is a desirable flavor. A manufactured synthetic sweetener could provide practical flavoring of these proteins to make them more palatable.

Mindful of the recent sugar shortage and of the increasing need to make available resources feed more people, man should think seriously about blending sugar and synthetic sweeteners to further the sugar supply. Corn and sorghum are cheaper and more abundant sugar sources than are cane and beets, but they produce glucose rather than sucrose. Pound for pound, glucose is as nutritive as sucrose, but it is only two thirds as sweet. Adding a synthetic sweetener to increase glucose sweetness would provide an immediate solution to the sugar shortage. Someday, man may have to consider "double-sweet" sugars to stretch the available carbohydrate. Someday, man may have to choose between stretching the food supply and triage.

Some practical synthetic sweeteners are known, and others are being developed. Needs exist for such products, and other needs will evolve. One of the greatest is the need for a more progressive governmental program which recognizes and permits use of these valuable food additives.

## REFERENCES


1. **Amerine, M. A., Pangborn, R. M., and Roessler, E. B.,** *Principles of Sensory Evaluation of Food,* Academic Press, New York, 1965, 28.
2. **Anon.,** How to cash in on the low-cal revolution, *Progressive Grocer,* 48(3), 180, 1969.
3. **Beck, K. M.,** Sweeteners, nonnutritive, , Encyclopedia of Chemical Technology, Vol. 19, 2nd ed. Kirk-Othmer, Ed., John Wiley & Sons, New York, 1969, 593.
4. **Beck, K. M., Jones, R. L., and Murphy, L. W.,** New sweetener for cured meats, *Food Eng.,* 30, 114, 1958.
5. **Beck, K. M. and McCormick, R. D.,** Current saccharin actions place added stress on need for a nonnutritive sweetener, *Food Prod. Dev.,* 6, 34, 1972.
6. **Beck, K. M. and Nelson, A. S.,** Latest uses of synthetic sweeteners, *Food Eng.,* 35, 96, 1963.
7. **Beck, K. M.,** Practical considerations for synthetic sweeteners, *Food Prod. Dev.,* 9(4), 47, 1975.
8. **Birch, G. G., Green, L. F., and Coulson, C. D.,** *Sweetness and Sweeteners,* Applied Science, London, 1971.
9. **Clauss, K. and Jensen, H.,** Oxthiazinone dioxides – a new group of sweetening agents, *Angew. Chem., Int. Ed. Engl.,* 12, 869, 1973.
10. **Coon, J. M.,** Naturally occurring toxicants in foods, *Food Technol.,* 23, 55, 1969.
11. **Daniels, R.,** *Sugar Substitutes and Enhancers,* Noyes Data Corp., Park Ridge, N.J., 1973, 3.
12. **Endicott, C. J. and Gross, H. M.,** Artificial sweetening of tablets, *Drug Cosmet. Ind.,* 85, 176, 1959.
13. **Farnsworth, N. R.,** Current Status of Sugar Substitutes, *Cosmet. Perfum.,* 88, 27, 1973.
14. **Helgren, F. J., Lynch, M. F., and Kirchmeyer, F. J.,** A taste panel study of the saccharin "off-taste", *J. Am. Pharm. Assoc. Sci. Ed.,* 44, 353, 1955.
15. **Horowitz, R. M. and Gentili, B.,** Taste and structure in phenolic glycosides, *J. Agric. Food Chem.,* 17, 696, 1969.
16. **Inglett, G. D.,** *Symposium: Sweeteners,* AVI, Westport, Conn., 1974.
17. **Lynch, M. J. and Gross, H. M.,** Artificial sweetening of liquid pharmaceuticals, *Drug. Cosmet. Ind.,* 87, 324, 1960.
18. **Moncrieff, R. W.,** *The Chemical Senses,* 3rd ed., The Chemical Rubber Co., Cleveland, 1967.
19. **Salunkhe, D. K., McLaughlin, R. L., Day, S. L., and Merkley, M. B.,** Preparation and quality evaluation of processed fruits and fruit products with sucrose and synthetic sweeteners, *Food Technol.,* 17, 85, 1963.
20. **Shallenberger, R. S. and Acree, T. E.,** Molecular theory of sweet taste, *Nature,* 216, 480, 1967.
21. **Wicker, R. J.,** Some thoughts on sweetening agents old and new, *Chem. Ind.,* No. 41, 1708, 1966.
22. **Miller, D. F. and Voris, L.,** Chronological changes in the recommended dietary allowances, *J. Am. Diet. Assoc.,* 54, 109, 1969.

Chapter 5

# APPLICATION POTENTIAL FOR ASPARTAME IN LOW CALORIE AND DIETETIC FOODS

**C. I. Beck**

## TABLE OF CONTENTS

# I. INTRODUCTION: OVERVIEW OF THE MARKET

The low calorie and dietetic foods market, in the U.S. alone, approaches 3 billion dollars annually. This growing market includes dairy products, soft drinks, confections, snacks, baked goods, canned goods, complete dinners, mix products, spreads, dressings, and seasonings. The majority of food items classed as "dietetic" are really "calorie-reduced" rather than therapeutic. That segment of the calorie-reduced foods market, in which the nondietetic counterparts derive a significant portion of calories from sugar, has a sales volume approaching 1 billion dollars at retail. Soft drinks comprise about half of this market, depending upon the regulatory status, quality, price relative to sugar, and public image of the commercially available high potency sweeteners.[1] High potency sweeteners can function in low calorie or dietetic foods by permitting calorie reduction or the removal of sucrose or both.

Most consumers select calorie-reduced foods to prevent, retard, or reverse obesity. Because about one fourth of the American population at any point in time is avoiding, limiting, or restricting food intake to maintain or lose weight,[2] low calorie foods appear to be a worthwhile objective as well as a sizeable market.

The reduction of sucrose intake in the diet has been indirectly recommended in several recent publications which relate the consumption rate of sucrose (about 100 lb/person/year in the U.S.)[3] to dental caries, coronary heart disease, hypertriglyceridemia, diverticular disease, diabetes, dermatitis, detrimental change in vision, and hypoglycemia.[4-8] Diverticular disease is ascribed to sucrose because refined sugars do not provide dietary fiber.[9-11] Many of these attacks on sucrose have gained momentum recently in both the scientific and popular press, but most issues are still highly controversial.[12,13]

Some persons view sucrose as a causative agent in diabetes,[14] but most medical experts are concerned with the relationship of sucrose to the dietary management of diabetes. Though the biochemical relationship between carbohydrate intake and serum insulin levels has long been studied,[15,16] control of total calories now seems to be more important in patient management than sucrose intake *per se*. On the basis of their research, the American Diabetic Association published revised dietary standards[17] in 1976.

Sucrose, as a major causative agent in dental caries, is well established and is not scientifically controversial.[18,19] There have been recommendations for reducing the cariogenic effects of sucrose by water fluoridation, restriction of sweets between meals, selection of liquid foods over solid foods, avoiding sticky and slow-dissolving sweets, brushing and flossing after eating meals or sweets (or at least rinsing the mouth with warm water after consuming any carbohydrate), seeking dental advice regularly, maintaining the teeth in an excellent state of repair, increasing research efforts to find anticaries food additives for use in foods containing sucrose and to establish the cariogenicity level of sucrose in different foods.[20] Though some of these suggestions are idealistic, there seems to be little question that their adoption would benefit caries-prone individuals and constructively influence national dental health. The American Dental Association has published a proposal which would require that the added sugar content in foods be stated in 5% increments on the labels of all formulated foods containing sugar. Giant Food, Inc., Washington, D.C., a major grocery chain, is considering an exploratory program in product labeling which would indicate added sugar in presweetened cereals.[21] A recent Gallup survey shows that 50% of a group of consumers polled have altered their food purchases because of concern about sugar content in processed foods.[22]

Other carbohydrates or even noncarbohydrate materials, if used to physically or functionally replace sucrose at similar levels, might also promote caries. Any substance being considered as a sugar substitute needs to be thoroughly studied in a controlled program similar to studies carried out at the National Institute of Dental Research, Bethesda, Maryland.

In addition to the preceding discussion of sucrose and dietetic foods, it seems appropriate to comment on the virtues of dietetic foods and on currently used marketing claims that can potentially damage the food industry. For a few products, such as chewing gum, it is reasonable and justified to claim and promote a "sugarless" image. The use of sweet food additives is necessary for the formulation of these products, but since natural foods are touted as being superior to processed foods by natural food freaks, the food industry apologetically claims in large print: "No cyclamates, No additives, No sugar, or Nothing Artificial." The use of these claims may discourage customers from using processed foods and further diminish an already limited food supply. Dietetic foods fulfill a need which is ethical and marketable. It would seem that acceptable products of good quality can be made and sold for their positive dietetic attributes. In a country that abounds with natural food freaks, it is discouraging to see established food industries reinforcing the folly of these uninformed idealists.

Though the removal of sucrose may have a basis in dietetic foods, the focus of this chapter is high potency sweeteners, especially aspartame, and related food technological issues. The following discussion, therefore, will emphasize the logic and specifics of preparing low calorie food products. It is the intention of this chapter to:

1. Discuss some considerations in formulating calorie-reduced foods
2. Provide specific formulation data to assist the bench technologist in preparing low calorie products
3. Present detailed data regarding aspartame, its successful applications, and its limitations
4. Establish that the preparation of low calorie foodstuffs involves many reformulation factors other than sugar replacement and that aspartame can make important contributions in this overall process
5. Stimulate further research on the efficacy of treating obesity with calorically reduced food products

## II. GENERAL CONSIDERATIONS FOR LOW CALORIE FOODS

Formulation of low calorie foods to affect weight loss presumes a great deal about the eating patterns and motivations of the obese, an area of study in which consistent large population data are lacking. Simplistically, it is assumed that people will substitute good tasting low calorie counterparts for traditional foods and will eat the same size portions, otherwise making no changes affecting weight. If this premise is accepted for consumption of low calorie foods, there are a number of factors to be considered when formulating low calorie products based on a high potency sweetner.

1. The percentage of calorie reduction for a serving of low calorie food equal to a serving of traditional food must be significant and must still comply with regulatory restrictions and standards. Though some foods have fewer calories per serving because air has been incorporated, most products can be significantly reduced in calories on a weight basis. This reduction usually will be accomplished by incorporating water or a low calorie bulking agent in the formulation or simply by reduction of fat content when compared to normal commercial counterparts. Watery products such as drinks having sugar as the prime source of calories are the products that can most easily be reformulated and still satisfy regulations.
2. The food selected for calorie reduction should have a high-caloric density.
3. The food selected for calorie reduction must contribute a significant proportion of calories to the diets of those who are overweight.
4. The food selected for calorie reduction, by using a high potency sweetener, must derive a high proportion of its calories from replaceable carbohydrate sweeteners. Note: Currently, there are no products or prospects for low or noncaloric protein substitutes. Calorie-reduced fats are being researched (see Section V).
5. The food selected for calorie reduction must be one in which the properties of substituted sugar can also be replaced. That is, sugar is also used as a humectant, a bulk material, a partner of the browning reaction, as well as a water-activity lowering agent for microbiological control. Unless a fully functional bulking agent becomes available, high potency sweeteners will have the greatest effect on calorie reduction in beverage mixes and bottled beverages and the least benefit in solid and semisolid foods such as baked goods and confections.
6. The food selected for calorie reduction should have a composition, flavor, storage life, and manufacturing process compatible with the physical properties of the sweetener to be used. It is important to realize that every system must be individually tested for suitability, despite any theoretical predictions.
7. In addition to reduced calories, the reformulated food must have good taste and high overall acceptability.

## III. CALORIE SOURCES

A puristic view of calorie derivation may lead one to the impractical conclusion that fats and oils should be removed from the diet. Calories obtained from food constituents are estimated as follows

| | |
|---|---|
| Carbohydrates | 4 cal/g |
| Protein | 4 cal/g |
| Fat | 9 cal/g |

Clearly, the elimination of fats from the diet (essential fatty acid and vitamin issues aside) removes a substantial portion of calorie contributions. Fats and oils are not consumed alone, but are absorbed as integral parts of food systems. Typical high fat foods are butter, chocolate, potato chips, nuts, and avocados. It is possible, with difficulty, to reduce fat content in chocolate and fried foods, but reduced intake of fat from butter, nuts, or avocados necessitates reduced consumption of these food products, though not necessarily abstention.

To identify sources of calories, several surveys by government agencies and private industry were studied in vain hopes of generating a table that would identify the specific food groups and their per capita caloric contribution to the average diet. Had this been possible, it still would have failed to show caloric contribution variation with sex, age, geographic region, race, or most importantly, weight class.

Information[23] detailing the consumption of available agricultural products has been obtained. Table 1 shows that approximately 16% of one's caloric intake originates from sugar and other sweeteners; this constitutes approximately 36% of the carbohydrate intake. If these data represent an average diet and if it were possible to replace all sugar with a noncaloric sweetener, only a 16% reduction in total diet calories would be possible. Clearly, high potency sweeteners are only one factor in a program for caloric reduction.

Based on the composition of the foods shown in Table 1, the fat, carbohydrate, and protein contribution of each class to overall energy intake can be determined. Stare[7] cites caloric contribution by food group to American diets as follows:

| | | |
|---|---|---|
| Total carbohydrates | 45% | Starch 30% |
| Fats | 40% | Sugars 15% |
| Protein | 15% | |
| | 100% | |

Research by Page and Friend[3] in 1972 shows that although carbohydrates contribute 45% of the calories, 47.2% of the total carbohydrate calories come from starch, and 52.8% originate from sugar; this is about 21% of the total calories from starch and 24% of the total calories from sugar. In either case, there appears to be an opportunity to substitute

TABLE 1

**Contribution of Major Food Groups Based on Food Supplies Available for Civilian Consumption[23]**

| | Calories (%) | Carbohydrates (%) | Fats (%) | Protein (%) |
|---|---|---|---|---|
| Meat, poultry, fish | 19.9 | 0.1 | 34.1 | 41.6 |
| Flour and cereals | 19.9 | 35.9 | 1.4 | 17.8 |
| Fats, oils, butter | 17.5 | 0.0 | 42.4 | 0.2 |
| Sugar and other sweeteners | 16.2 | 35.9 | 0.0 | 0.0 |
| Dairy products | 11.4 | 6.7 | 13.2 | 22.0 |
| Fruits | 3.5 | 7.3 | 0.4 | 1.1 |
| Beans, peas, nuts, soy | 3.1 | 2.3 | 3.7 | 5.6 |
| Vegetables | 3.0 | 5.5 | 0.5 | 3.9 |
| Potatoes | 2.9 | 5.6 | 0.1 | 2.4 |
| Eggs | 2.0 | 0.1 | 3.0 | 5.0 |
| Coffee, chocolate Equivalents, miscellaneous | 0.7 | 0.5 | 1.3 | 0.4 |
| Total | 100.1 | 99.9 | 100.1 | 100.0 |

*Note:* Data based on 1975 preliminary figures.

TABLE 2

**Sugar Deliveries by Industry – Calendar Year 1975[24]**

| Consumption uses | Annual estimated U.S. per capita[a] consumption of sugar (lb) |
|---|---|
| Household use[b] | 22.4 |
| Beverages | 18.8 |
| Bakery, cereal, allied | 11.1 |
| Confectionery, etc. | 7.1 |
| Canned, bottled, and frozen foods, jams, jellies, preserves | 6.5 |
| Ice cream, dairy products | 4.5 |
| Hotels, restaurants, institutions[c] | 0.7 |
| Other food deliveries | 4.4 |
| Pharmaceutical, tobacco, other nonfood | 0.8 |
| Total (from identifiable food classes) | 76.3 |
| Deliveries to wholesale and retail grocers | 28.7 |
| Grand total | 105[d] |

[a]Assuming 215 million people.
[b]Deliveries in less than 50-lb units (all other figures delivered as hundredweights).
[c]Reported substantially larger consumption in 1971.
[d]Does not include government unspecified or bulk shipments.

between 15 and 24% of the food energy in the diet by sugar replacement.

Additional information shows how sugars and other sweeteners (Table 1) are ultimately utilized in industries producing products for home or institutional use.[24] Table 2 identifies product groups in which sugar use is significant. Analysis of these product groups identifies specific products in which replaceable sugar contributes substantially to the calories of the foodstuff.

## IV. REGULATORY LIMITATIONS

### A. Ingredients

Once marketing and research identify a food concept as a logical target for calorie reduction, a number of regulations must be studied before development work begins. First, it is necessary to determine what approved ingredients are available. The April 1975 issue of the Code of Federal Regulations (CFR) cites the safe use of saccharin salts as food additives on an interim basis (21 CFR Sec. 121.4001) and saccharin labeling requirements (21 CFR Sec. 125.7). Ammoniated glycyrrhizin is listed as a generally recognized as safe (GRAS) substance under essential oils, oleoresins, and natural extractives. Aspartame is cited as an approved food additive (21 CFR Sec. 121.1258) prior to a stay of that order published in the *Federal Register*, December 5, 1975 issue.

If a bulking material is used as a food ingredient, the section of the regulation dealing with "label statements relating to non-nutritive constituents" applies. It requires a label statement of the percentage by weight of the bulking agent, and in juxtaposition with the name, the term "non-nutritive" (21 CFR Sec. 125.7). This same regulation describes saccharin as a nonnutritive substance and "shall not be construed as authorizing the use of saccharin or its salts in any food other than one for use by persons who must restrict their intake of carbohydrates . . . ." This regulation seems to restrict use of mixed sweetener systems that employ both saccharin and carbohydrate sweeteners. Regulatory matters relating to sweeteners have been reviewed by Becker,[25] and Ebert has discussed development of new food additives.[26]

### B. Dietary Products

The general regulation entitled "Label statements relating to certain food used in control of body weight or in dietary management with respect to disease," must also be considered. This regulation (21 CFR Sec. 125.6) states that percentages of protein, fat, and available carbohydrate, as well as available calories per unit, must appear on the label. The mention of even one of these on the label of a nonspecial dietary product might require the entire nutrition labeling regulation (21 CFR Sec. 1.17), which was promulgated later, to be invoked. Regulations dealing with special dietary foods have been under review and are likely to be changed and expanded in the future. The terms "artificial sweetener, nutritive, and non-nutritive sweetener" are defined in 21 CFR Sec. 125.1 and 21 CFR Sec. 121.1 (o).

Since calorie-reduced products are classified with special dietary foods, regulations dealing with "General label statements; dietary properties; value; placement," (21 CFR Sec. 125.2) should be studied to avoid mislabeling due to implied or stated claims.

### C. Specific Product Approvals

It is necessary to study all regulations relating to a specific target product. Though the use of the terms diet, calorie-reduced, or low calorie imply a 50% reduction in calories when compared to a commercial nondiet counterpart, no such general regulation exists. There are specific diet beverage regulations (21 CFR Sec. 3.72) that permit manufacture of a diet product if it is at least 50% reduced in calories when compared to a sugar sweetened beverage and if its caloric density does not exceed 6 cal/fluid ounce.

An example of a regulation which expired from nonuse is one that deals with canned fruits (21 CFR Sec. 3.78). This regulation allowed the manufacture of diet canned fruit products when calories were reduced at least 50% from the heavy syrup counterparts. Where 50% reduction was not technologically feasible, a product more than 25% reduced in calories could have been approved as "diet" by special petition to the Food and Drug Administration (FDA).

In both the diet beverage and diet canned fruit citations, the use of nonnutritive and nutrive sweeteners was specifically allowed in mixtures, whereas 21 CFR Sec. 125.7 was inconsistent with these two regulations. (Note: A recent proposal published in the Federal Register (July 19, 1977) would define a low calorie food as having 40 calories or less per serving (and 0.4 cal/g or less) and a calorie reduced food as having less than 2/3 of the calories of a nonobscure counterpart.)

### D. Standards of Identity

Finally, before commencing product development, it is critically important to determine what formulation restrictions exist if the target food category is standardized. Ice cream is an example of food which demonstrates the importance of regulatory limitations.

A typical ice cream has a composition of 12% milk fat, 20 to 24% milk solids (including the milk fat), and 12 to 15% sugars (sucrose or mixtures with corn syrup). It would have about 1.7 lb/gal of total digestible food solids and about 4.6 lb total weight/gal in the finished product. According to the U.S. Department of Agriculture Handbook No. 8, this product will have about 207 cal/100 g.

Present standards for ice cream (21 CFR Sec. 20.1) specify minimum milk fat, minimum milk solids, optional sweeteners excluding nonutritive sweeteners, total digestible food solids controlling the addition of water and/or bulking materials, and total weight of the finished product controlling the overrun-air. Since the product must have at least 4.5 lb/gal of finished product and at least 1.6 lb/gal must be digestible food solids, then 36 g/100 g of product must be digestible. Since the product must also be 10% milk fat, that leaves 10 g of fat and 26 g of carbohydrate and protein. Thus, the product has a minimum of 194 cal/100 g. The maximum caloric reduction is only 6%. When a nonnutritive sweetener is to be used in ice cream, it must be petitioned for use as an optional ingredient. The likelihood for approval of its use is uncertain.

The International Association of Ice Cream Manufacturers is participating in revision of these standards and has developed a set of proposed regulations that are unlikely to have an impact on development of a legal definition for "diet" ice cream.

Is there a satisfactory resolution between what the food industry wants and what current regulations allow? Using sophisticated technology, acceptable products can be made at 110 cal or less per 100 g; however, they cannot be called ice cream or ice milk. Creative marketing will determine if these products can compete without a standardized identity. By deviating from standards, a great deal of flexibility is gained in formulation, product placement, distribution, and processing.

It is interesting to note that the FDA recently issued regulatory letters to ice cream manufacturers in upper New York State to stop the production of frozen dietary dairy desserts containing sorbitol and saccharin.[27] Objection to the product is based on the premise that saccharin is approved only for caloric reduction. Since frozen dietary dessert is formulated for diabetics, saccharin was not seen to serve a valid special dietary purpose.

**E. State and Local Regulation**

In addition to these concerns with federal regulations, it is critically important to determine what state and local regulations apply to the product class under development. State offices and city boards of health should be consulted, as they often serve specific local regulatory functions.

## V. THE NEED FOR BULKING AGENTS

Table 2 shows the end product uses of sucrose as estimated from reported deliveries of sugar in hundredweights. There are no available data concerning end uses of household sugar. The following constitute probable use: in hot and cold beverages, on fruits and cereals, in baked goods, in icings and glazes, in home preserves, and in a wide variety of smaller volume uses such as puddings, whipped toppings, salad dressings, confections, sauces, and candied fruits and vegetables.

In soft drinks, dry mix beverages, or tabletop sweeteners for hot or cold beverages, direct substitution for sugar by a high potency sweetener is uncomplicated. However, when the sugar in cake is replaced with a high potency sweetener, the product is no longer cake (see Figure 1). Sugar functions as more than sweetening in a cake. Commercial cakes have a sugar level providing optimum cake volume. Table 3 shows sugar properties that must be acknowledged when using a high potency sweetener to replace sugar in solid or semisolid foods.

To replace the volume of removed sugar, a bulking agent must serve some of the same functions as sugar, but it should not have the calories of sugar. It may or may not be sweet or have preservative action. Other foods that contain either a high percentage of sugar or are solid or semisolid foods include cookies, candy, frozen desserts, chewing gum, pudding and pudding mix, frosting mix, and forms of high intensity sweeteners for table use. Food consumption data for 1967 indicate that the market volume of a bulking agent needed to accomplish a 1% penetration into the sugar market for calorie reduced substitutions is about 36 million lb/year.[28]

If a bulking agent must be indigestible to be low- or noncaloric, a number of likely bulking

TABLE 3

**Properties of Sucrose Which an Ideal Bulking Agent Must Possess**

| |
|---|
| Texture |
| Bulk |
| Compatability |
| Stability |
| Noncrystallizing in certain foods |
| Solubility |
| Tasteless, other than sweet |
| Colorless |
| Humectant properties |
| Preservative (optional) |
| Sweetness (optional) |
| Caloric (must be low- or noncaloric) |

FIGURE 1. Effect of sucrose on cake structure and volume.

agents can be identified from these categories: cellulose derivatives, hydrocolloids (gum), polyhydric alcohols, synthetic polymers, sucrose-esters, and silica gels. Many of these are highly viscous. When replacing large quantities of sucrose, they generate a cement-like casting of the product in which they are used. Others may cause diarrhea when consumed in large amounts. Some are regulated for uses in which they are limited to low percentage levels. As of October 1976, no genuine broad usage bulking agent was approved in the U.S. Several are under development: polydextrose[29] (Pfizer Inc., Groton, Connecticut); maltitol[30] (Hayashibara Biochemical Laboratories, Inc., Okyama, Japan); and Palatinit[31] (Sueddeutsche Zucker-Akiengesellschaft, Mannheim, W. Germany). Other nonnutritive materials such as dietary fiber are being developed, but the physical properties of sucrose are not claimed.

Some work has been done with functional replacements for fats and oils, although much additional development and biological testing will be required if these compounds are ever to become food additives. Researchers at Proctor & Gamble, Cincinnati, Ohio, have been studying sucrose polyester compounds[32,33] with an average of six to eight of the hydroxyl groups esterified. Their studies show that these compounds, having fat-like properties in food applications, are not absorbed,[34] are safely handled at 50 g/day without causing diarrhea in humans,[35] and furthermore lower cholesterol absorption.[36] Other researchers have studied compounds that may have application as noncaloric or low calorie fats.[37-39]

## VI. HIGH POTENCY SWEETENERS

A thorough search of the literature reveals hundreds of sweet compounds; however, at this writing, the use of only two intense sweeteners is permitted in foods in the U.S. – saccharin and ammoniated glycyrrhizin. Both have serious drawbacks in taste. Saccharin is well known for its bitter aftertaste, and the glycyrrhizin products (even at low sweetness levels) contribute a lingering sweetness and aromatic licorice taste to foods. The licorice taste of glycyrrhizin so limits its applications that for all intent, saccharin is "the only game in town."

A number of substances have been blended with saccharin to reduce the aftertaste and off-flavor. Daniels[41] reviews the patent literature for the blends claimed to reduce saccharin bitterness. Compounds that can be satisfactorily blended with saccharin include dipeptides, glucono delta lactone with buffers or sodium gluconate, calcium gluconate, lactose, galactose, ribonucleotides, piperazine, pectin, mannitol/gum, maltol, adipic acid, and sodium chloride. A glycine and saccharin blend is approved for beverages (21 CFR Sec. 121.1257). The lactone, gluconate, and lactose additives are utilized in sweeteners for table use, and adipic acid is used in dry-mix products. Though saccharin is currently permitted for use in dietary foods and beverages, its regulatory future is uncertain. The U.S. General Accounting Office (GAO) has been very critical of the FDA's handling of saccharin, stating that the sweetener should not be allowed to remain on the market under an interim regulation.[40]

Two other sweeteners, aspartame and cyclamate, have received considerable publicity in the past few years. Both have had marketing approval for use in foods, but neither was being marketed as of December 1977. Cyclamate is not as intense a sweetener as saccharin, but also has off-taste and lingering taste.[42] When saccharin and cyclamate were available, they were blended at about 10 parts cyclamate to 1 part saccharin to maximize taste acceptability. Salant has reviewed applications of this combination at 12:1 ratios.[43] Cyclamate use was banned in the U.S. in 1969. In 1976, the FDA upheld its earlier decision by refusing a Food Additive Petition for use of cyclamate submitted by Abbott Laboratories, North Chicago, Illinois. Aspartame is a high potency sweetener with probably the most sugar-like taste of sweeteners discovered to date. Although aspartame has many applications in sweetening foods and beverages, it has some limitations because of instability in hot and/or neutral to alkaline conditions. These limitations might be compared with those of volatile or heat-sensitive flavoring compounds, where judicious technology has overcome application limitations. A number of other peptide sweeteners have been cited.[44] In late 1975, the FDA stayed the regulation for aspartame pending further investigation of G. D. Searle & Co.'s support data for aspartame.[44a] To date, a Public Board of Inquiry, which will be convened by the FDA to review objections to the original 1974 regulation, has not yet been scheduled.

Another cluster of substances with intense sweetness might be classed as the "exotic" sweeteners. Included are extracts of miracle fruit (*Synsepalum dulcifium*), stevioside (*Stevia rebaudiana*), wild red berry (guinea potato or Serendipity berry, *Dioscoreophyllum cumminsii*), katemfe or the miraculous fruit of Sudan (*Thaumatococcus daniellii*),[45] and the Jerusalem artichoke.[46] The Jerusalem artichoke (sunflower root) should not be confused with globe artichoke (*Cynara scolymus*). The extract of the Jerusalem artichoke is not sweet in itself; it develops its sweetness when heated or processed with food products. The *cumminsii* berry extract contains a protein which has been named monellin by researchers who isolated it at the Monell Chemical Senses Center, where it was originally isolated. The katemfe sweetener is also proteinaceous and is called thaumatin. The miracle fruit extract (miraculin) is not really a sweetener in itself. When miraculin is consumed, the taste buds become treated, and any acidic product consumed within a few hours tastes sweet. Unfortunately, liquids such as beer and wine consumed hours after initial exposure to miraculin have also been reported to taste overly sweet.[47] The Miralin Company (formerly of Hudson, Massachusetts) attempted to commercialize this product under the name of Miralin®. However, in September 1974, the FDA decided to remove miraculin from the GRAS list and reclassify it as a food additive. The Miralin Company could not raise the money to appeal FDA's decision or to undertake comprehensive tests needed to win clearance, and hence fell to bankruptcy.

There are many other sweet substances described in the literature. For some time, Horowitz and Gentili[61,62] at the U.S. Department of Agriculture (USDA) and others, have pursued the development of the dihydrochalcones. These compounds, derived from citrus flavanones, have high intensity, but lingering sweetness. In 1969, Nutrilite Products Inc., Lakeview, California, filed a limited Food Additive Petition for use of neohesperidine dihydrochalcone in chewing gum, but was asked to conduct additional safety testing still in progress. Researchers at Stanford Research Institute, Menlo Park, California, have been studying some oxime compounds related to perillartine, having potential value as sweeteners. However, concerns have developed with hydrolysis products including hydroxylamine. Hoechst, Frankfurt, W. Germany developed an easily soluble intense sweetener under the generic name of acesulfam and initiated the necessary testing in preparation for clearing of its use in foods. Hoechst's studies are expected to be complete in 1978. Eli Lilly and Co., Indianapolis, Indiana, has developed two sweeteners: 6-chloro-D-tryptophan and melizame, 5(3-hydroxyphenoxy)-1*H*-tetrazole. In addition to resolving enzyme inhibition, some research will be required to establish that the D-tryptophan derivative does not interfere with normal metabolism.

Xylitol,[48] a carbohydrate sweetener obtained from wood, corn cobs, and other cellulose substances, has a sweetness level near that of sucrose. Although claimed to be noncariogenic, xylitol has about the same caloric content as sucrose. It is currently approved for limited dietary uses abroad. However, FDA is considering revoking the regulation of 1971, based on reports of adverse effects in humans.[8] (Note: Recently (November 17, 1977), xylitol was implicated as a carcinogen by the FDA based on data from Huntington Laboratories in England.)

Many extensive structure and activity studies have been done on sweeteners. Studies reported in the literature[49-51] have produced numerous theories. Sweeteners today are still being discovered serendipitously, or by empirical modification of structures already known to be sweet. There are some general review references discussing the spectrum of sweeteners.[41,52-55] Using these and primary literature, an attempt has been made to expand the available tabular data. The data are shown in Table 4; but no attempt was made to include sweeteners with potency less than 20 times the sweetness of sucrose on a weight basis.

## VII. A TECHNICAL VIEW OF ASPARTAME

### A. Historical Aspects

Aspartame was inadvertently discovered as an intermediate during the synthesis of a product for ulcer therapy in December 1965.[58] Upon the discovery of aspartame, a broad program of dipeptide synthesis ensued,[44] and ultimately the project came full circle as the aspartame dipeptide was selected for commercialization.

Aspartame is white crystalline powder with almost no odor and with a clean, sweet taste generally 180 to 200 times the sweetness of an equal weight of sucrose in foods. It has sufficient solubility for most applications, but is excluded

TABLE 4

**Status and Qualities of High-Potency Sweeteners[a]**

| Compounds | Comments | Ref. |
|---|---|---|
| **Current usage (U.S.)** | | |
| Ammoniated glycyrrhizin | A triterpene derivative; 50X sucrose; MacAndrews and Forbes Co., Philadelphia, Pa.; isolated from licorice root (*Glycyrrhiza glabra*); substantial licorice taste at normal use levels | 41, 56 |
| Saccharin | 2,3-dihydro-3-oxobenzisosulfonazole; 300–500X sucrose; Sherwin-Williams Co., Cleveland, Oh.; bitter aftertaste at normal use levels; confirmation of safety was under review by the GAO, and continues by FDA. | 40, 41, 43 |
| **Current interest** | | |
| Acesulfame-K | 6-methyl-1,2,3-oxathiazin-4(3 K)-one-2,2-dioxide; 130–200X sucrose; Hoechst Aktiengesellschaft, Frankfurt, W. Germany; 3-year safety studies in West Germany should be completed during 1978 | 57 |
| Aspartame | L-aspartyl-L-phenylalanine methyl ester; 150–200X sucrose; G. D. Searle; Skokie, Ill.; taste is clean and free of significant off-flavors; unstable in highly acidic or basic systems; heat labile; FDA approval granted in 1974, order stayed in December 1975 due to questions of validity of safety data; Searle is actively seeking reapproval | 58, 59, 100 |
| Cyclamate | Cyclohexylsulfamate; 30–40X sucrose; Abbott Laboratories, N. Chicago, Ill.,; some metallic aftertaste; removed from U.S. market by FDA order in 1969; order reaffirmed in mid-1976; Abbott is actively seeking reapproval | 43, 60 |
| Hesperetin dihydrochalcone glycoside | Dihydrochalcone derivative of de-rhamnosyl citrus peel flavanone glycoside; 80–290X sucrose; flavor less lingering, more fruity than neohesperidine dihydrochalcone | 45, 61, 62 |
| Melizame | 5(3-hydroxyphenoxy)-1H-tetrazole; 170X sucrose; Eli Lilly & Co., Indianapolis, Ind.; slight delay or lingering of sweetness; sucrose-like flavor; company has indicated a continuing interest and is actively considering a clearance program | 63, 64 |
| Monellin | 91 amino acid residue protein; mol wt 11,700; 2000X sucrose; Monell Chemical Senses Center of the University of Pennsylvania and the Veterans Administration Hospital, Philadelphia, Pa. and Unilever Research Labs, Hamburg, Germany; source is berries of *Dioscoreophyllum cumminsii*; sweetness can be persistent; heat labile, acid stable | 65–67 |
| Naringin dihydrochalcone | Dihydrochalcone derivative of a citrus peel (seville orange) flavanone glycoside; 50–300X sucrose; Nutrilite Products (subsidiary, Amway Corp., Lakeview, Calif.) some menthol aftertaste with salty or slightly bitter notes; toxicology studies in progress | 45, 61, 62 |

[a]Some of the more recent information was obtained by personal communication.

TABLE 4 (continued)

**Status and Qualities of High-Potency Sweeteners[a]**

| Compounds | Comments | Ref. |
|---|---|---|
| Neohesperidine dihydrochalcone | Dihydrochalcone derivative of a citrus peel (grapefruit) flavanone glycoside; 250–1800X sucrose; Nutrilite Products Subsidiary, Amway Corp., Lakeview Ca.) and California Aromatics & Flavors, Palo Alto, Ca.; licorice-like aftertaste; slow sweetness onset; little bitterness; performs best alone in chewing gum, mouthwash, toothpaste and as a blend in other products; most soluble of the dihydrochalcone sweetners; well-advanced toxicology program | 45, 61, 62 |
| Oxime V | 4-(methoxymethyl)-1,4 cyclohexadiene-1-carboxaldehyde *syn*-oxime; 450X sucrose; Stanford Research Institute, Menlo Park, Ca.; hydrolytic stability claimed satisfactory for soft drink applications above pH 3; a modification of Perillartine; may cost less than cyclamate for equal sweetness; only acute and subacute safety data generated to date; compound lacking the 4-(methoxymethyl)-substituent is also patented as a sweetener | 66, 68, 69 |
| Paramethoxycinnamaldehyde | Usage range is 0.0005 – 0.3% by weight; Proctor & Gamble Co., Cincinnati, Oh.; patents indicate primary usage in mouthwashes and toothpastes | 70, 71 |
| Perillartine | An oxime of a monoterpene aldehyde; 2000X sucrose; synthesized from a component of the essential oil of *Perilla frutescens*; approved in Japan since 1920; used in tobacco; bitter with menthol-licorice aftertaste; poor solubility | 52, 66, 69, 72, 73 |
| 3-amino-4-n-propoxybenzyl alcohol | Dynapol, Palo Alto, Ca., development no longer being pursued | 74 |
| **Uncertain status** | | |
| L-alanine *tert*-butyl ester | 35–40X sucrose | 75 |
| 2-(N-aspartylamino)-alkanol fatty acid esters | 120X sucrose; Tanabe Seiyaku K. K., Osaka, Japan; claimed to be stable in aqueous solution | 76 |
| L-aspartylaminomalonic acid methyl fenchyl diester | 33,000X sucrose; Takeda Chemical Industries Ltd., Tokyo, Japan; stability in food systems may limit applications; the methyl 2-methylcyclohexyl diester has a reported sweetness of 6600X sucrose | 77–79 |
| L-aspartyl-L-(β-cyclohexyl)-alanine methyl ester | 130–240X sucrose; G. D. Searle & Co., Skokie, Ill. | 41, 59, 80, 81 |
| L-aspartyl-L-homoserine methyl ester | 80X sucrose; Kyowa Hakko Kogyo K. K., Tokyo, Japan | 82 |
| Chlorinated sucrose | 1′,4,6,6′-tetrachloro-1′,4,6,6′-deoxygalactosucrose; 500–600X; claimed to have no licorice or bitter aftertaste | 83 |
| 4β-10α-dimethyl-1,2,3,4,5,10-hexahydrofluorene-4α-6-dicarboxylic acid | Diterpene derivative; 1300–1800X sucrose; derivative of a major component of pine rosin; 10X as bitter as caffeine | 84 |
| Heliotropyl nitrile | Piperonyl nitrile; usage range is 0.001–20.0% by weight; Proctor & Gamble Co., Cincinnati, Oh. | 85 |

TABLE 4 (continued)

**Status and Qualities of High-Potency Sweeteners[a]**

| Compounds | Comments | Ref. |
|---|---|---|
| 5-imino-4-dimethyl-2-imidazolidinone | 100–200X sucrose; Shell Oil Company, Houston, Tex.; no undesirable aftertaste | 41, 86 |
| Ionic sweetener | Synthesized dihydrochalcones; 400–700X sucrose; Dynapol, Palo Alto, Ca.; improved aftertaste compared to other dihydrochalcones; development continues | 87 |
| Miraculin (Miralin®) | Glycoprotein; mol wt 44,000; normal amino acid distribution; not a sweetener; a taste modifier which can make acidic foods taste sweet for 1–2 hr after tongue contact with 1–2 mg; developer was Miralin Corp., Hudson, Mass., applied for GRAS status and denied; source is fruit of *Synsepalum dulcificum* (Schum. and Thonn.) Baehni | 65, 66 |
| *N*-trihaloacetyl aspartic acid dipeptide esters and anilides | 120–3000X; American Home Products Corp., New York, N.Y.; the anilide sweeteners are the more potent of this group | 88 |
| Nonabsorbable polymer-bound sweeteners | A class of 4-alkoxy, 5-hydroxy (or 5-amino), 1-polymer-linked nonabsorbable compounds; Dynapol, Palo Alto, Ca.; development continues | 66, 89 |
| Thaumatin I, II (Talin®) | I: 193 amino acid residue protein; mol wt 21,000; II: 186 amino acid residue protein, mol wt 20,400; some question if they are distinct entities; 1600X sucrose; Unilever Research Labs, Hamburg, Germany; acid and heat labile; extract of katemfe, miraculous fruit of Sudan (*Thaumatococcus danellii*); Tate and Lyle, Ltd. (London, Eng.) is developing thaumatin as Talin®; 90-day feeding studies are completed, the company is approaching the U.S. and Canada for required additional testing | 65, 90 |
| **Development unlikely** | | |
| Chlorogenic acid | 3-caffeoylquinic acid; a 2,3 dihydroxycinnamic acid monoester of quinic acid; not a sweetener; a taste modifier causing perception of sweetness in foods following its ingestion; extract of heart of globe artichoke | 91 |
| Cynarin | 1,5 dicaffeoylquinic acid; see chlorogenic acid | 91 |
| Dulcin® | 4-ethoxyphenylurea; 100–300X sucrose; now considered toxic | 45 |
| *N*-acetylkynurenine | 35X sucrose; good flavor, immediate impact, and quick dissipation; ultimate manufacture considered difficult | 92, 93 |
| *N*-formylkynurenine | 35X sucrose; bitter off-flavor due to hydrolysis to formic acid on the tongue; ultimate manufacture considered difficult | 92, 93 |
| Lo Han Ko sweetener | Triterpenoid glycoside; 150X sucrose; General Foods, White Plains, N.Y.; delayed and lingering sweetness; stable at 100°C in aqueous solution; approval would require structure elucidation and approval of source, Lo Han Fruit (*Momordica grosvenori*) | 94 |

TABLE 4 (continued)

**Status and Qualities of High-Potency Sweeteners[a]**

| Compounds | Comments | Ref. |
|---|---|---|
| Osladin | A steroid saponin; 3000X sucrose; extract of rhyzome of *Polypodium vulgare* L. | 95 |
| P–4000 (Ultrasweet®) | 1-*n*-propoxy-2-amino-4-nitrobenzene; also called Ultrasweet®; 4000X sucrose; toxic, nitrobenzene can damage blood-forming tissues; at least 15 related nitrobenzenes have been reported; sweetness range 25–2000X sucrose | 55 |
| Stevioside | A diterpene extracted from the leaves of *Stevia rebaudiana*; 300X sucrose; safety is unresolved – the related steviol and dihydrosteviol have evidenced antiandrogenic activity | 60, 96, 97 |
| D-Tryptophan and derivatives | Enantiomer of a naturally occurring essential amino acid; 25–40X sucrose with 6-chloro-, 6-trifluoromethyl-, 6 methyl-, and 6 fluoro-derivatives ranging from 600–1300X sucrose; Eli Lilly & Co., Indianapolis, Ind.; metabolic studies contraindicate use as sweetener; effects include irreversible inhibition of a liver enzyme by a metabolite of the 6-chloro compound | 98, 99 |

*Note:* X means compound is (No.) times sweeter than sucrose on a weight basis.

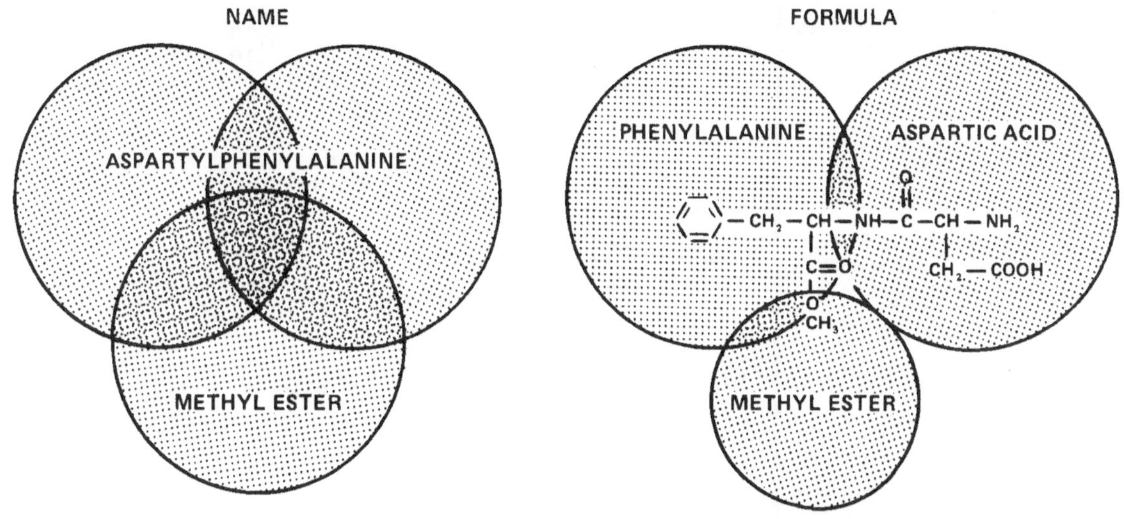


FIGURE 2. Chemical structure of aspartame.

from some wet systems where heat, high pH, and time combine to convert the aspartame to its tasteless diketopiperazine and/or consequently to the amino acids. An enormous amount of work has been done[136] to establish the safety of both aspartame and its diketopiperazine. Several general articles have been written describing aspartame's properties,[59,100–102] and its generalized use in foods is described in the patent literature.[103]

## B. Chemistry

Chemically, aspartame is the esterified dipeptide 1-methyl-*N*-L-α-aspartyl-L-phenylalanine shown in Figure 2. The chemical formula is $C_{14}H_{18}N_2O_5$, and molecular weight is 294.3 Its synthesis and analysis are described by Mazur et al.[104] Being a peptide, the compound is amphoteric, and the dissociation constants at 25°C are $pK_1 = 3.1$ and $pK_2 = 7.9$. The isoelectric point (minimum charge and minimum solubility) is at pH = 5.2.

Solubility is a function of pH and temperature as shown in Figure 3. Note that maximum solubility is at pH = 2.2. Though minimum solubility can go as low as 0.5%, that concentration would be equal to nearly 100% sugar when one recalls the sweetness intensity of aspartame. Solubility in nonaqueous solvents is shown in Table 5. Note that aspartame is insoluble in vegetable oil; this insolubility can prevent or retard the conversion of aspartame to its diketopiperazine. Ariyoshi has also described the relative solubilities of the α- and β-L-aspartyl-L-phenylalanine methyl esters in a wide range of solvents.[105]

Prolonged storage in aqueous solution, exposure to heat, pH conditions outside of 3.0 to 5.0, or combinations of these factors can cause the aspartame molecule to convert to its diketopiperazine, to the dipeptide, and ultimately to the amino acids as shown in Figure 4. Conversion to the diketopiperazine is more likely at higher pH values, and peptide cleavage to amino acids prevails at acid pH values. The stability of aspartame is adequate for most food systems, though care must be taken with processing conditions. A discussion of aspartame inclusion in specific food systems follows.

The analysis of aspartame in foods is particularly complex since aspartame is an esterified dipeptide. Peptides, proteins, and protein hydrolyzates as well as esters (primarily from fruits and flavors) are ubiquitous, and background interferences must be handled on a product-by-product basis. Published methodology is presently limited to analysis in pure form or in model systems.[106,107]

## C. Manufacturing

The manufacture of aspartame is complicated by the fact that the sweet isomer is the α form of the L-L-dipeptide ester. The β-dipeptide is reported to be bitter,[105] and the D-L-, L-D- and D-D-dipeptides are not sweet. Thus, the basic process is one of preparing both L-amino acids, the anhydride of L-aspartic acid, the methyl ester of

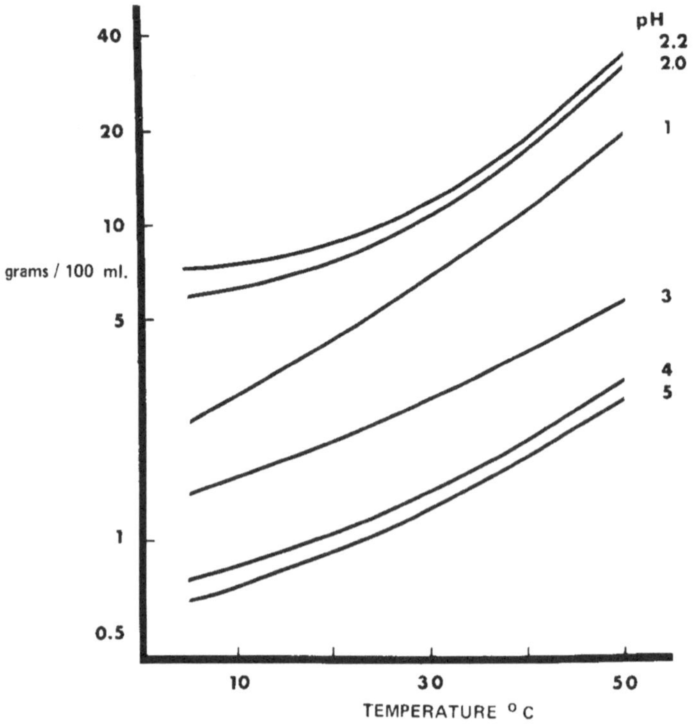


FIGURE 3. Effect of pH and temperature on aspartame solubility in water.

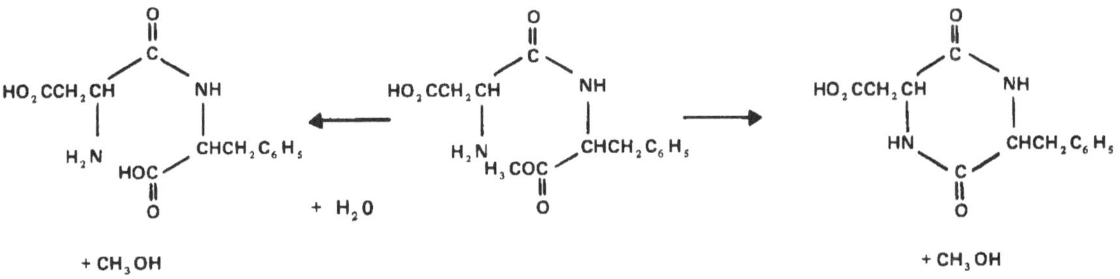


FIGURE 4. Reactions for conversion of aspartame (center) into its diketopiperazine (right) or its unesterified dipeptide (left).

TABLE 5

**Solubility of Isoelectric Aspartame at 25°C**

| Solvent | Solubility (%) |
|---|---|
| Water | 1.0 |
| Methanol | 0.9 |
| Ethanol | 0.4 |
| Chloroform | 0.03 |
| Heptane | 0.004 |

L-phenylalanine, and then coupling the anhydride to the methyl ester to give the α-form, aspartame. The manufacturing technology for production and purification is described in the patent literature.[108,109] An alternative process has also been reported.[110]

### D. Sweetness Quality

The clean sweetness of aspartame was cited earlier and, in addition to a substantial base of experience and industry response, was formalized via flavor profiling techniques.[59] Cloninger and Baldwin, in their study of sweetness in various foods, comment on lack of off-flavor in aspartame systems even at sweetness levels equivalent to 10% sucrose.[111,112]

The most extensive work in taste quality was presented by Pangborn and Larson and documented by Larson.[42] Time-intensity figures were shown in which aspartame is very similar to sucrose regarding time-intensity relationships for flavor, bitterness, and sourness in distilled water; in three flavors of still drinks; and in two flavors of gelatin. The time-intensity curves for sweetness did quantify earlier observations that aspartame's sweetness is slightly persistent. Saccharin and cyclamates were included in this study and generally proved stronger and persistently more bitter. Descriptive analysis, which compared each of the three noncarbohydrate sweeteners to sucrose, showed responses to a wide range of descriptive terms for aroma, flavor, and aftertaste in the same products. In general, sucrose- and aspartame-sweetened samples were described as "sweet clean." Cyclamate and saccharin samples were considered "sweet chemical" and "bitter." Moskowitz and Dubose studied taste qualities of aspartame and mixtures containing sugar.[113]

The difference in taste quality between saccharin-sweetened commercial diet soft drinks and soft drinks prepared with aspartame sweetener in an identical fashion will be discussed in Section VIIIC, on beverages. Panelists preferred aspartame-sweetened diet soft drinks (in which the sweetness was drastically reduced) to commercially available saccharin products. In a clinical study (discussed later), covert substituted products were not perceived as calorie-reduced by a small group of obese subjects participating in the program.

Though aspartame's taste properties have been generally quite acceptable in most food systems, some researchers have reported methods for further improvement. Various mixtures containing glucono-delta-lactone, gluconate salts, cream of tartar, sodium bicarbonate, lactose, and glucose are said to yield improved sweetening properties.[114,115] Aluminum potassium sulfate, naringin, or a combination of these two (when used with aspartame) is said to control and reduce a lingering sweet taste.[116]

### E. Flavor Enhancement

Though aspartame is effective predominantly as a sweetener, it can also influence, potentiate, or enhance other flavors or their perception. Proving this, like proving synergy, is quite involved due to the need to separate the effect of added sweetness from other concomitant flavor changes. Aspartame in chewing gum compositions extends sweetness to *keep* "the flavor on the bedpost overnight."[117] Whereas sugar gum compositions lost flavor in 5 min, added aspartame doubled the flavor retention in one example and lasted seven times longer (35 min) in sugar formulations where greater amounts of aspartame were added. When compared to a saccharin formulation, an almost identical chewing gum made by substituting aspartame for saccharin in a 2:1 ratio had detectable flavor for six additional minutes.

Additional work has indicated that aspartame can reduce the bitterness of low grade coffee at levels that are not perceived as sweet.[118] Aspartame is credited with improving the taste and aroma of fruit-flavored drinks and gelatin desserts prepared from aspartame-sweetened mixes. From 0.001 to 0.02% of aspartame by weight was required to achieve this effect.[119]

### F. Sweetness Intensity

The sweetness intensity of aspartame, though variable in systems, is generally about 180 to 200 times that of an equal weight of sucrose when both are isosweet.[59] This potency can change in a given product as a result of minor formulation changes in flavor or viscosity.[112,120] Small amounts of gelatin or methyl cellulose effectively raise the perceived sweetness of aspartame in a buffered aqueous solution.

Table 6 shows aspartame's sweetness intensity in a wide range of applications. A number of sweetness evaluations in water have been included because they demonstrate the variability of findings, even in the least complicated system.

Specific data dealing with aspartame are also available; aspartame is synergistic with saccharin,[59,111,121,122] cyclamate,[111,122] sucrose,[123] and glucose.[123] The magnitude of synergy in these blends is substantial. Though synergy experiments with saccharin are difficult because of the interfering bitterness, blends of aspartame and saccharin at a 3:2 ratio on a weight basis, in a formulated tabletop sweetener, have a synergy in hot coffee of about 50% based upon assumed potencies of 300 for saccharin and 200 for aspartame. Similar synergy for that blend has also been found in soft drinks. This means that only

TABLE 6

**Reported Potency Values for Aspartame in Various Media**

| Medium | Potency[a] | Sucrose in the control medium (%) | Tasting temperature (°C) | Ref. |
|---|---|---|---|---|
| Water | 400 | 0.34 | 22 | 59 |
| | 215 | 4.3 | | |
| | 133 | 10 | | |
| | 100 | 15 | | |
| | 182 | 2 | 8 | 112 |
| | 160 | 4 | | |
| | 100 | 12 | | |
| | 43 | 30 | | |
| | 61 | 10 | 3 | 42 |
| | 54 | 10 | 22 | |
| Buffer-citrate | 160 | 4 | 8 | 112 |
| Phosphate pH 3.2 | 100 | 12 | 8 | |
| Beverages | | | | |
| Powdered mixes | 133 | 10 | 8 | |
| | 143 | 10 | 3 | |
| | 180 | 11 | 8 | |
| Soft drinks | 180 | 10 | 8 | 59 |
| Coffee[b] | 180 | 4 | 70 | |
| Tea,[b] hot | 180 | 6 | 70 | |
| Tea,[b] iced | 180 | 7 | 8 | |
| Fruit drink | 180 | 14 | 8 | |
| Orange juice Frozen concentrate Undiluted | 280 | 34 | Chilled | |
| Chewing gum | 160–195 | 69 | 22 | 117 |
| Desserts | | | | |
| Vanilla frozen dessert | 225–300 | 15 | Soft frozen | |
| Pudding | 200 | 17 | 8 | 59 |
| Gelatin – Flavor 1 | 160 | 18 | 8 | |
| Flavor 2 | 172 | 18 | 3 | 42 |
| Flavored yogurt | 175–220 | 17 (13 added) | 8 | |
| Cereals | | | | |
| Presweetened | 180 | 37 | – | |
| W/spoon-for-spoon | 220 | Variable | – | 59 |
| Fresh fruit[b] | 220 | Variable | – | |
| Canned fruit | 175 | 20 | Chilled | |
| Confections | | | | |
| Marshmallows | 70–120 | 57 | Ambient | |
| Chewable lozenge | 70 | 91 | Ambient | |

[a]Potency is equal to weight of sucrose divided by weight of aspartame at equal sweetness.
[b]Spoon-for-spoon equivalent product used as described in applications.

two thirds of the calculated blend is required to perform the sweetening function.

## G. Sweetener Mixtures

The blending of sweeteners has long been known to accentuate the best of each sweetener.[124-129] Cyclamate and saccharin blends have a higher quality of sweetness than either does alone, and the synergy of the compounds can result in significant cost savings. Synergy simply means that 2 + 2 > 4. Percent synergy is the excess divided by the expectation; if 2 + 2 = 5, then 1 ÷ 4

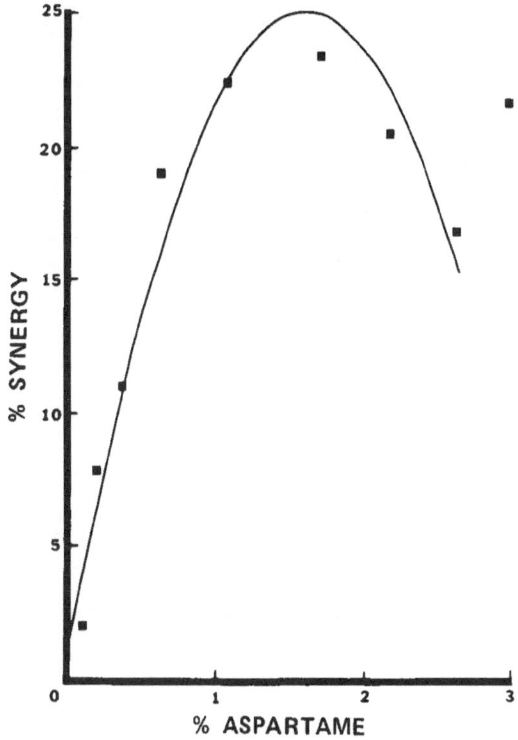


FIGURE 5. Sweetness synergy of mixtures containing sucrose and aspartame, in lemonade at a sweetness level equal to 10.2% sucrose.

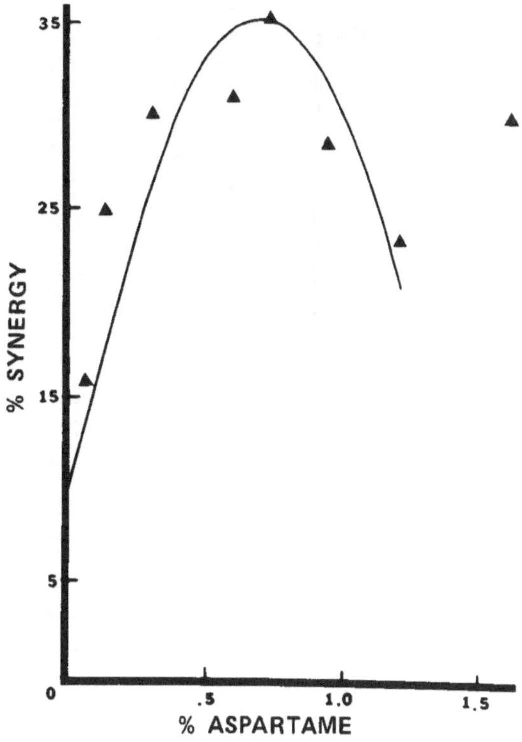


FIGURE 6. Sweetness synergy of mixtures containing glucose and aspartame, in lemonade at a sweetness level equal to 10.2% sucrose.

= 25%. An example in the literature shows that a solution of 0.053% sodium saccharin also containing 0.533% sodium cyclamate would, by calculation, have the sweetness of a 10.8% sucrose solution (4.3% sugar for the saccharin and 6.5% sugar for the cyclamate). The blend is actually sweeter than a solution containing twice the calculated percent sucrose.[130]

If nutritive and nonnutritive sweeteners can be blended, similar benefits will accrue; maximum synergy with sucrose is 23% at 1.7% aspartame in the blend and with glucose; it is 35% at 0.72% aspartame in the blend. These sensory data were obtained in a lemonade system at the sweetness level of 10.2% sucrose. It was tasted at about 12°C; approximately 70 subjects participated in each test. The relationships of blend composition and percent synergy are shown for aspartame with sucrose, with glucose, and with saccharin (see Figures 5 to 7).

### H. Stability

To maximize the performance and stability of aspartame in food systems, it is necessary to understand each of the parameters affecting its stability. Moisture, temperature, pH, and time play important roles in conversion of aspartame to methanol and its dipeptide or its diketopiperazine (3-carboxymethyl-6-benzyl-2,5-diketopiperazine).

Dry stability of aspartame is quite good. Samples of sachet products packaged in standard food-service paper packets have been kept in normal warehouse storage for over 3 yr and still meet a 95% α-aspartame specification. The effect of temperature on the pure, dry ingredient can be measured easily at temperatures far exceeding normal handling (see Figure 8). In a dry environment, on thermogravimetric analysis, 196°C is required to form the diketopiperazine; for decomposition to occur, 315°C is necessary.

Since dry stability is shown to be far better than aqueous stability, it is quite important to know at what percent moisture the product ceases to be "dry." Products were prepared at various moistures (4.2, 5.5, 6.0, 6.5, and 8.0%) and sealed in glass ampuls. They were stored at room temperature, 40 and 55°C for 1 year. No sign of

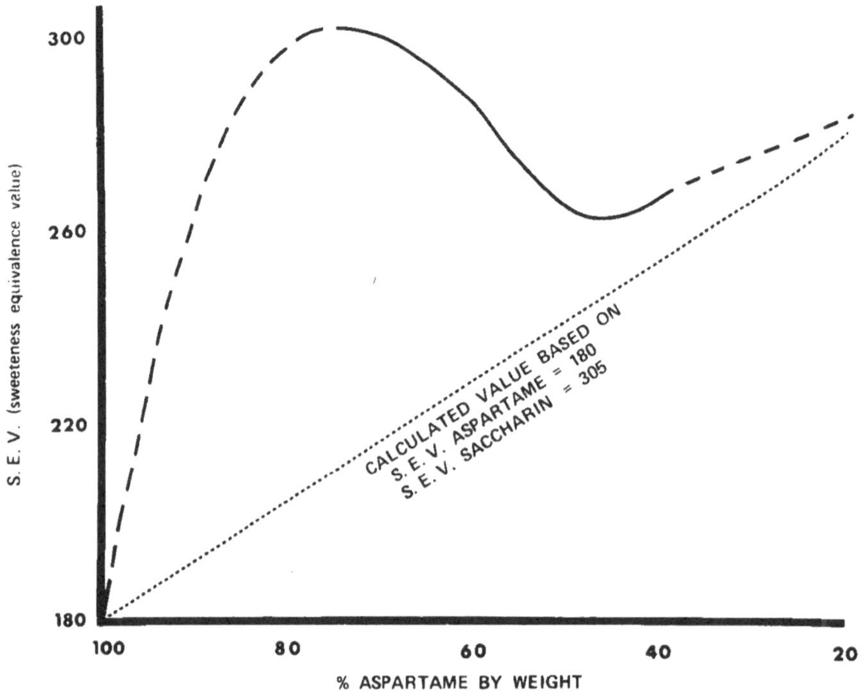


FIGURE 7. Sweetness synergy of mixtures containing saccharin and aspartame. Solid lines are experimental; dashed lines are extrapolated based on theoretical expectations.

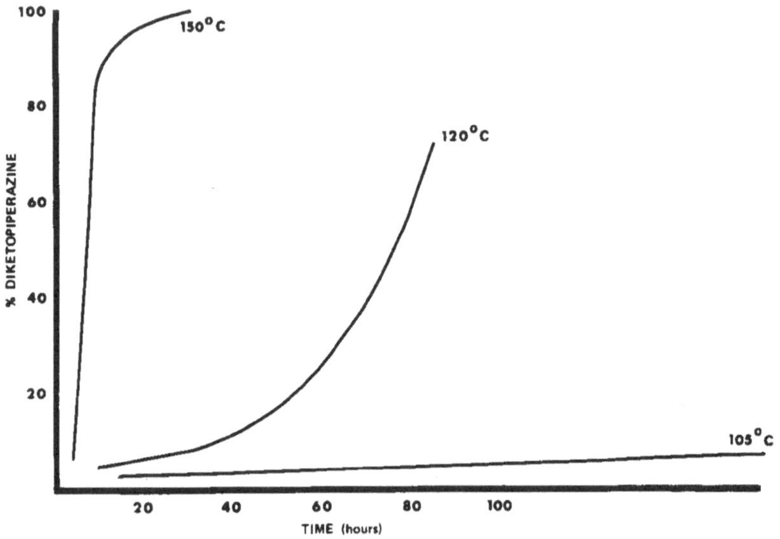


FIGURE 8. Stability of aspartame in dry storage.

decomposition was seen below 8% moisture. At 8% moisture and 55°C, 2.5% decomposition occurred at 26 weeks; similarly at 40°C for 52 weeks, 2.2% decomposition occurred. The only sample kept at 55°C for 52 weeks was almost totally decomposed, and additional information at higher moisture levels is unfortunately unavailable. Either the one ampul was wetter than 8%, or high temperature decomposition is autocatalytic. The important conclusion is that aspartame is dry even under severe heat, up to almost 8% moisture, and possibly further.

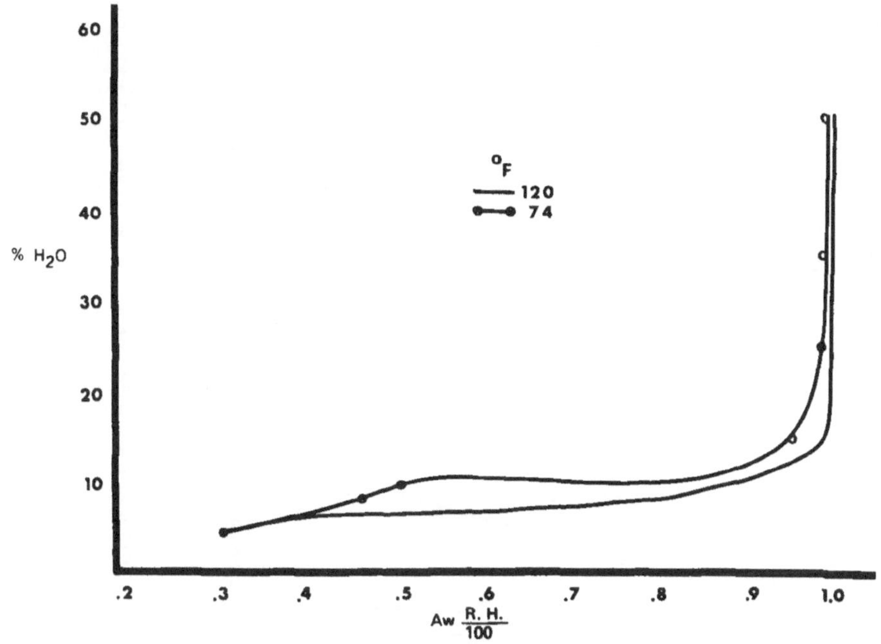


FIGURE 9. Moisture equilibrium isotherm for aspartame.

If one is concerned with keeping aspartame dry, it is consequently necessary to consult the moisture equilibrium isotherm shown in Figure 9. This would indicate that even up to 85% relative humidity, aspartame would barely reach 12% moisture at equilibrium. This should provide latitude in the selection of packaging materials for dry mix products, but the moisture vapor properties and moisture of other ingredients in a blend or co-dried product must not be overlooked. Carbohydrates and/or acidulants have been proposed as fusion mediums into which aspartame can be encapsulated to stabilize it from moisture.[131]

Within wet systems, temperature, time, and pH all are important considerations. Fortunately, aspartame has its greatest stability in the weak acid range in which most foods exist, pH 3.0 to 5.0. At pH 3.4 and lower, hydrolysis to the dipeptide becomes most important; above pH 4.8 to 5.0, the additional cyclization to its diketopiperazine (DKP) causes a dramatic reduction in sweetness. Figure 10 shows the stability of aspartame at 40°C in aqueous buffer at pH values from 1 to 8; a horizontal line shows complete stability, whereas a vertical line indicates instantaneous conversion and loss of sweetness. Figures 11 to 13 show the same information for 55, 68, and 80°C, respectively. Cursory consideration of pH alone might mislead the technologist to think that dairy products, eggs, and neutral vegetables are poor applications. A cold milk amplifier, dry mix blender shake, hot chocolate mix, or frozen dairy dessert are all tangible formulations because of short time exposure and/or the low temperatures of these products.

Temperature is also quite important to the development of an aspartame-sweetened food. Table 7 shows conversion time for an aqueous product to reach 20% of its original aspartame when formulated at pH 4.0. In viewing this table, it is important to know something about loss of sweetness as consumers may perceive it. Research was done in the food laboratories at G. D. Searle & Co. to identify the least noticeable differences in sweeteners' used in familiar product systems like coffee and lemonade. Experienced but untrained panelists participated. To obtain a difference at the 95% confidence level, a 9% change in sweetener concentration was required for a paired comparison presentation. Significance with a single stimulus presentation required a 16% change in sweetener concentration.

Another reality that affects an integrative perspective is that flavors of commercial stored products vary. How these changes affect acceptability is important. In the following discussion of applications, data show that the rate of analytical disappearance of aspartame is rarely paralleled by loss of acceptance. The nonlinearity of the sweet-

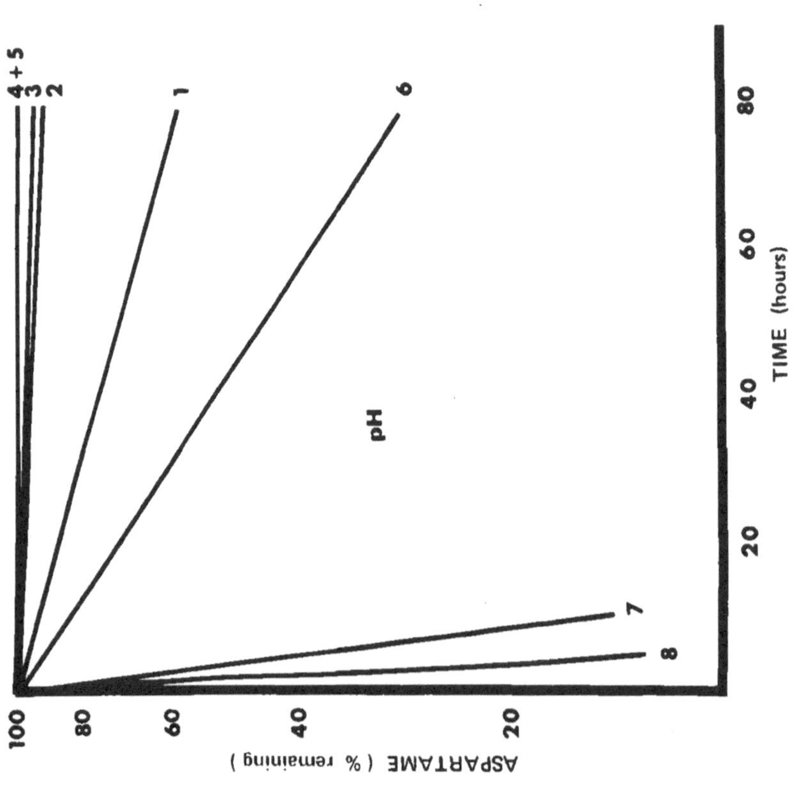


FIGURE 10. Effect of pH and time on aspartame stability in aqueous buffer systems at 40°C.

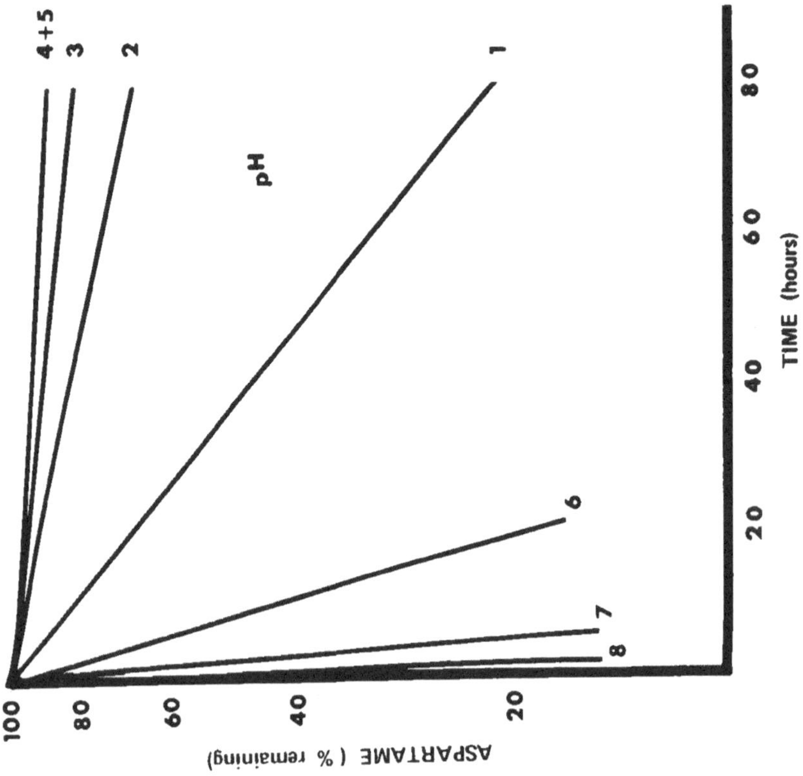


FIGURE 11. Effect of pH and time on aspartame stability in aqueous buffer systems at 55°C.

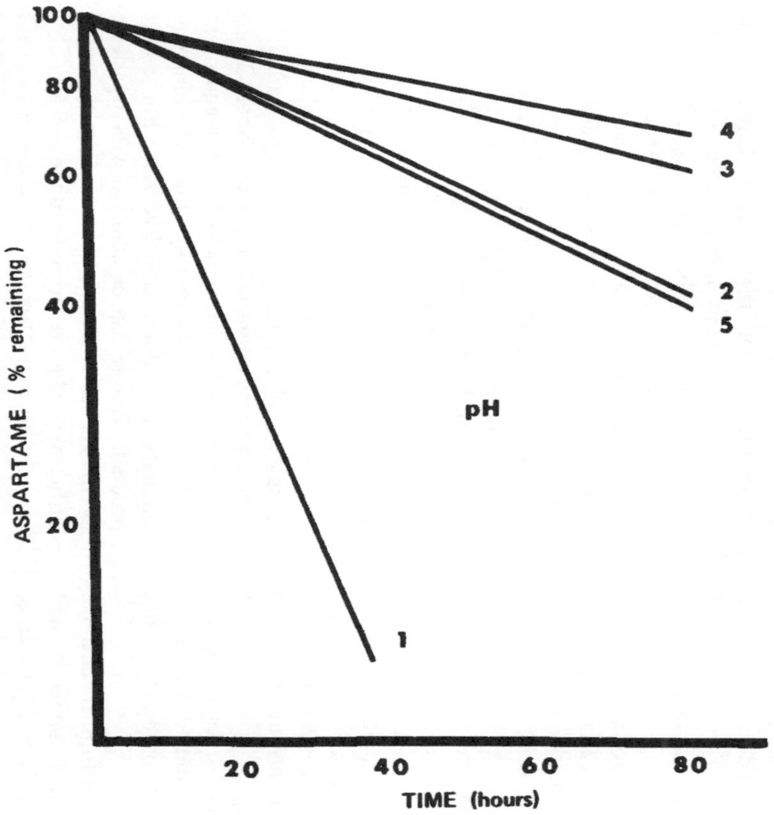


FIGURE 12. Effect of pH and time on aspartame stability in aqueous buffer systems at 68°C.

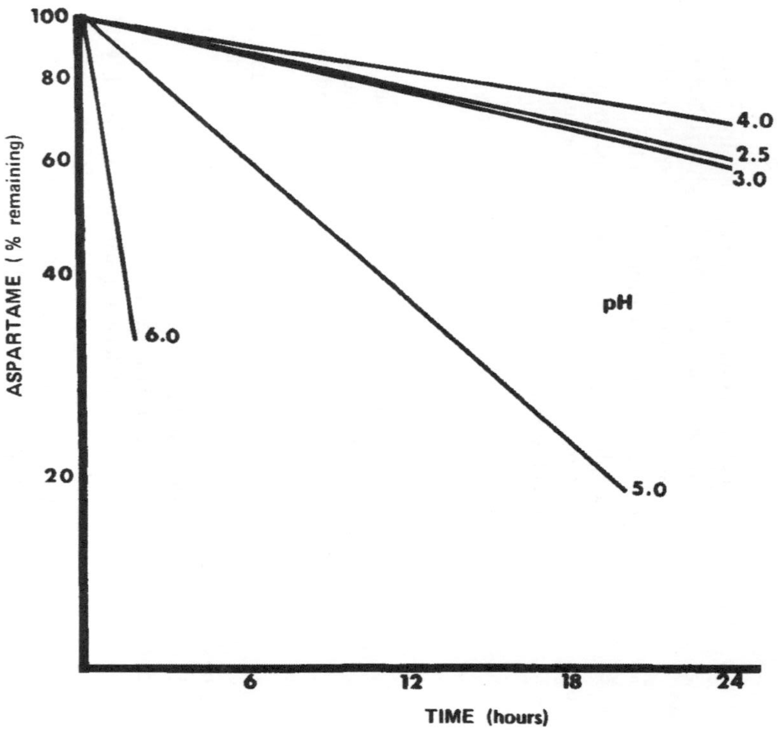


FIGURE 13. Effect of pH and time on aspartame stability in aqueous buffer systems at 80°C.

TABLE 7

**Effect of Storage Temperature on Aspartame Stability in Aqueous Solutions at pH 4.0**

| Temperature storage (°C) | Calculated time for 20% decomposition (days) |
|---|---|
| 10 | 387 |
| 20 | 134 |
| 30 | 51 |
| 40 | 22 |
| 55 | 5 |
| 68 | 2 |
| 80 | 1 |
| 90 | 0.15 |

ness response may partially explain this phenomenon. When aspartame is converted to its diketopiperazine, the only taste effect is sweetness loss due to loss of aspartame. The diketopiperazine has a negligible impact on taste.[59]

Lastly, the stability data presented here are primarily based on buffer model systems which, for unclear reasons, result in a faster aspartame loss than occurs in foods. The data shown here are accurate and the general trends indicated are true; however, each food system should be evaluated in its own stability test. The data presented here should be valuable in designing those tests.

### I. Metabolism and Medical Implications

In the course of evaluating aspartame as a new food ingredient and consequently preparing a Food Additive Petition, a massive amount of animal and clinical data have been assembled describing the effects of the sweetener and its diketopiperazine.[132-136] Aspartame, a peptide, is metabolized by the same biochemical pathways as proteins. Naturally, it has 4 cal/g. However, since so few grams are required (due to its sweetness intensity), it functions quite effectively in calorie reduction. The comparative manufacture and metabolism of aspartame, sugars, cyclamate, and saccharin are shown in Figure 14. Aspartame has undergone testing at the National Institutes of Dental Research where it failed to promote caries in rats.

Aspartame is broken down into phenylalanine, aspartic acid, and methanol. If all presently used sugars could be replaced by aspartame, the average aspartame consumption would be about 0.8 g/day, based on sugar consumption of 102 lb/yr. This provides a possible 5% increase in daily consumption of aspartic acid and for phenylalanine an increase of 10%. Because about 1 in every 10,000 individuals is born without the ability to metabolize phenylalanine, they should be aware that aspartame is a phenylalanine source, as are certain protein foods.

## VIII. FOODS FORMULATED WITH ASPARTAME

As discussed above, both therapeutic and marketing needs will be best met if low calorie foods are formulated in categories representing large sugar consumption. Therefore, the sequence of this discussion will generally follow that of Table 2.

### A. Household Use

*1. Table Top Sweeteners*

Intense sweeteners must be combined with ingredients that act as diluents before they are suitable for household use. These sweeteners can be prepared in different forms: tablets, liquid dispensers, packets (sachets), or as a granular spoon-for-spoon equivalent. Presently, only the liquid systems provide dilution without calorie contribution from the dilutent. However, the liquids raise concerns about microbiological hazards, portability, stability, shipping weight, and uniform dispensing. Add to this list aspartame's limited stability in aqueous systems, and it is clear that liquid dispensers are the least desirable product.

Both effervescent and noneffervescent tablets can be formulated. The noneffervescent tablets dissolve slowly, even in hot coffee. The effervescent tablets contain only about one third aspartame by weight. All tablets suffer from discontinous metering; one and three fourths of a tablet cannot be realistically measured. Deciding the optimum sweetener content per tablet is a complex problem that was considered earlier;[59] aspartame was optimized in coffee at about 20 mg per tablet. Sachet and spoon-for-spoon equivalent products meter sweetness by dilution in minimum amounts of caloric carbohydrates. For sachet products, lactose can be admixed with aspartame by inexpensive mechanical blending to assure sufficient product recovery from the package.

Moisture characteristics of the carrier can have an important effect on the selection of packaging material. Electrostatic interactions should not be

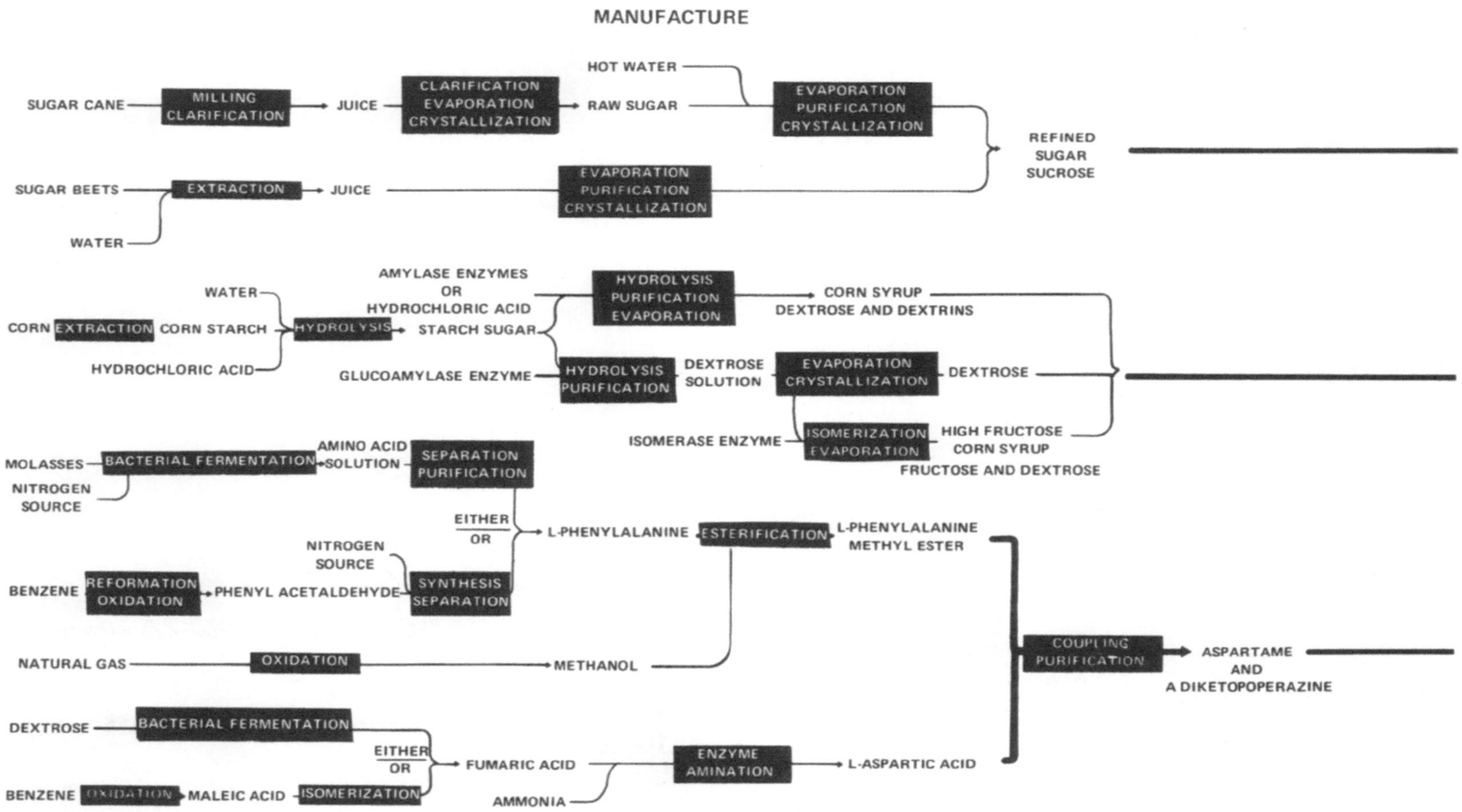


FIGURE 14. Comparative synthesis and metabolism of sugars, cyclamate, saccharin, and aspartame.

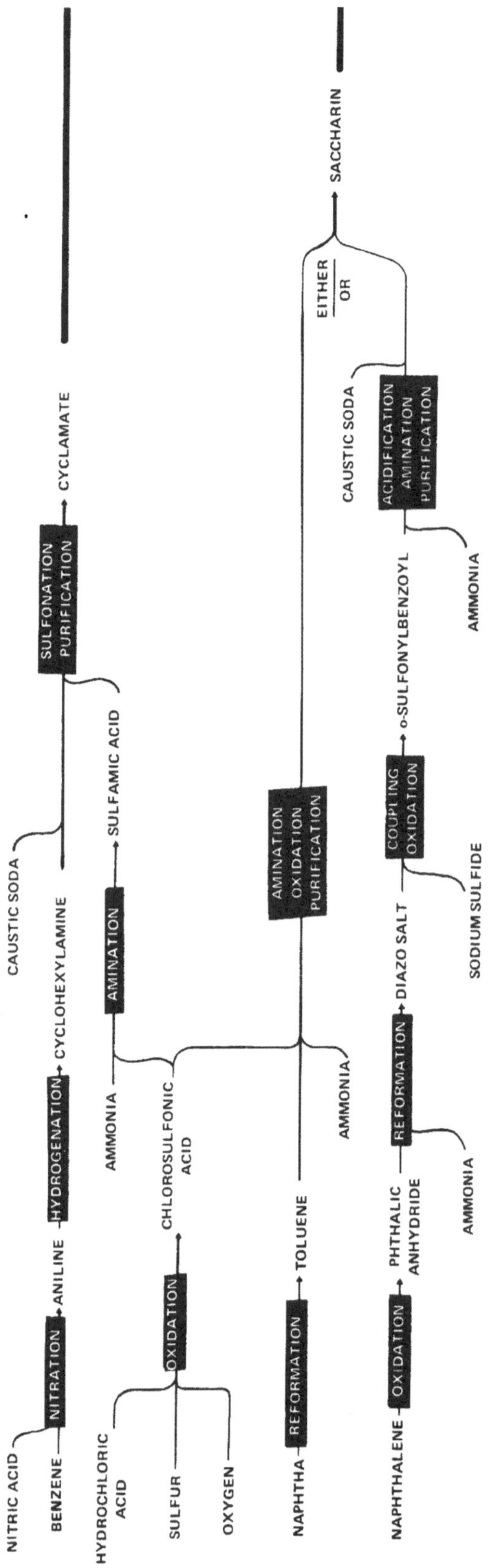


FIGURE 14 (continued)

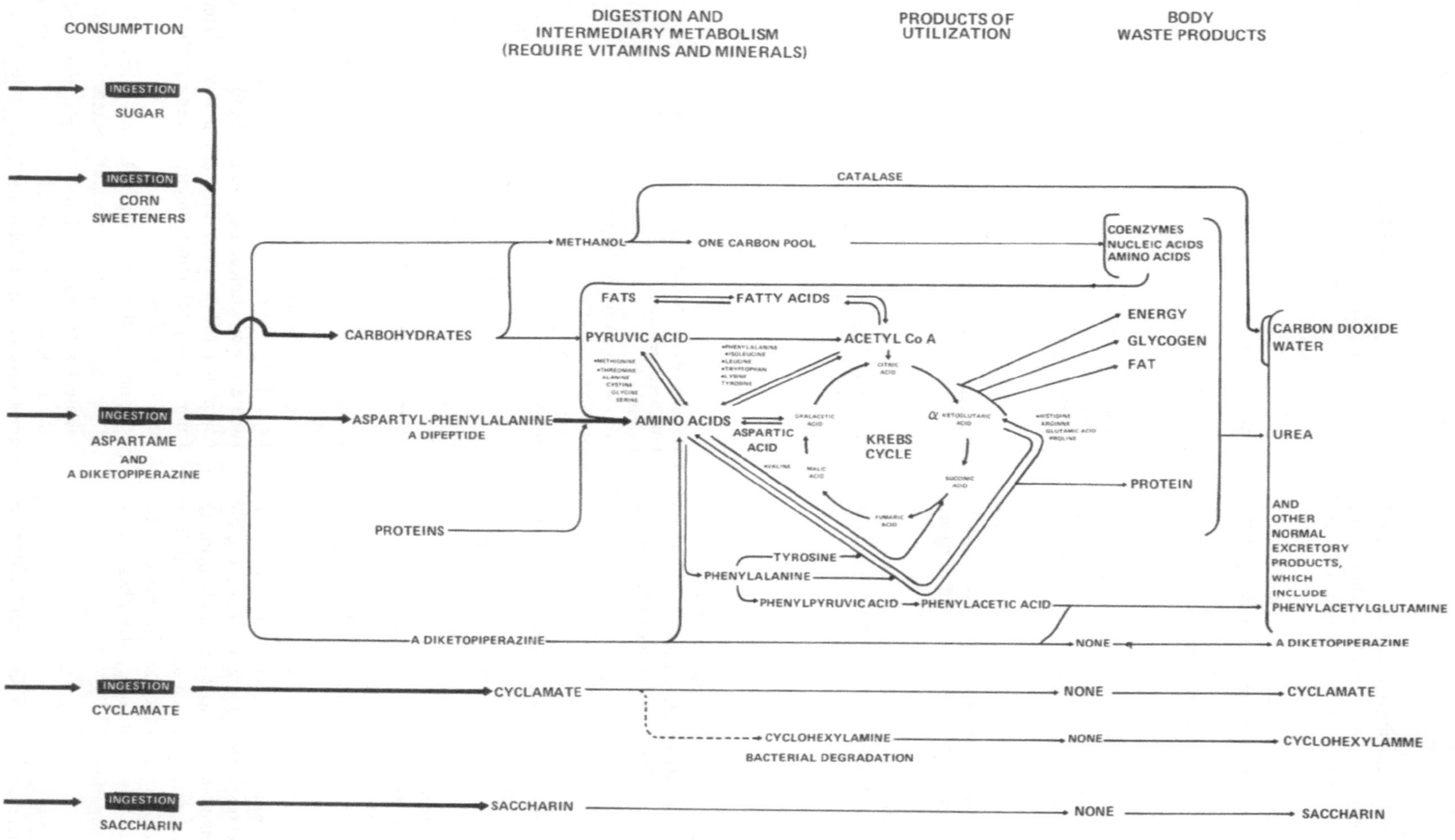


FIGURE 14 (continued)

overlooked. In the spoon-for-spoon products, the content of carbohydrate carrier is minimized by fluffing. The carrier is co-spray dried with the sweetener to a minimum practical density; very low density products might easily spill from a spoon. Spray-drier conditions will affect the degree of sugar-like visual perception of the product. Table 8 shows typical formulations for the sweetener's granular forms.

In tablets, lactose is used as a filler; bicarbonate and acidulant provide effervescence, and L-leucine in an amount not to exceed 3.5% acts as a lubricant to release the tablet from the press (CFR Sec. 121.1258). In the sachet, the lactose acts as a filler. In the spoon-for-spoon product, the Malto-Dextrin acts as a carrier and filler, while tricalcium phosphate acts as an antistatic agent. The surfactant reduces foam in processing prior to spray drying and reduces froth formation when the sweetener is dissolved in hot beverages. The spray-dried product has some added benefits. It dissolves faster in breakfast cereal and adheres more uniformly to fruit surfaces than does sugar.

Because of the osmotic effects of sugar, sweetening fruit with sugar is quite different from sweetening fruit with a high potency sweetener. A high potency sweetener such as aspartame protects the fruit texture, while sugar draws out juices and forms a syrup.

*2. Technology*

The formulation and production of granular tabletop products has been documented in the patent literature. The preparation of low calorie, readily soluble, low hygroscopic sweetening compositions is accomplished by preparing aspartame solutions with various low dextrose equivalent starch hydrolyzates, and then co-drying,[137] or more specifically spray-drying[138] the solutions. Acidulants are suggested for increasing solubility of the final product.[139,140] Additional improvements in the solution rate occur when the acidulant and sweetener are co-ground in an organic solvent in which aspartame is insoluble prior to drying.[141] Aspartame can also be formed in a matrix material composed of saccharides or food acids.[142] Still other technology describes stable, soluble aspartame forms.[143-145]

*3. Usage Patterns*

Just as there is little published information on sugar uses in the home, there are few data on the uses of low calorie sweeteners. A study was conducted in spring 1976 by G. D. Searle & Co. to determine use patterns for low calorie sweeteners. Four hundred adult female family members participated in a 4-week study; all had used artificial sweeteners in the 6 weeks prior to the study. Selected families had approximately four members. They were not diabetic and used artificial sweeteners by choice. Two hundred and forty-two completed the process. This group was selected from an original questionnaire screening 10,000 families (see Table 9).

Basically, the study indicated that about 40% of present granular saccharin-based tabletop sweeteners are used in hot beverages, and about 10% are used on breakfast cereal. Other uses were quite diverse. Approximately 90% of total consumption was by individuals older than 21 yr, and

TABLE 8

**Tabletop Sweeteners**

| Spoon-for-spoon product[a] | Concentration (%) |
|---|---|
| Spoon-for-spoon[a] | |
| Malto-Dextrin | 96.44 |
| Aspartame | 3.45 |
| Tricalcium phosphate | 0.1 |
| Surfactant | 0.01 |
| | 100 |
| Concentrated sachet[b] | |
| Lactose | 97.78 |
| Aspartame | 2.22 |
| | 100 |

[a]Spoon-for-spoon (spray-dried) density = 0.11g/cm³.
[b]Mechanical blend; 1 packet = 1 g = 2 tsp sugar.

TABLE 9

**Consumer Usage Patterns of Saccharin-based Tabletop Sweeteners in Homes Where Artificial Sweeteners Are Regularly Used**

| | Average usage (%) | |
|---|---|---|
| Usage category | Packet | Bulk |
| Hot beverage (coffee, tea, cocoa) | 44.2 | 33.1 |
| Cold beverage | 0.7 | 1.1 |
| Cereal | 11.4 | 9.1 |
| Other additive usage (sprinkled) | 5.7 | 8.0 |
| Ingredient in baking | 1.2 | 3.3 |
| Ingredient in stovetop cooking | 0.9 | 2.8 |
| Miscellaneous | 36.0 | 42.7 |
| | 100.0 | 100.0 |

approximately 1% was by children age ten or under. About two thirds of all consumption was by female homemakers. The analysis in Table 9 does not include tablets, presumed to be used almost entirely in hot beverages.

*4. Recipe Development*

It is necessary to educate the consumer in the most effective use of any new product. In the case of tabletop sweeteners utilizing aspartame, it is important to extend the household use as much as possible without suggesting abusive use with fried, baked, boiled, or broiled systems. Traditional and obvious uses in hot and cold beverages and on cereal and fruit were included. Home recipes were developed for salads and dressings; entrees such as "chicken and citrus wine sauce," and "sweet 'n sour meat balls"; desserts including gelatin, custard, and compote; and snacks such as cereal squares and cold-set oatmeal cookies. In most recipes, it was possible to add the aspartame after the heating step. Preparation of chocolate oatmeal cookies (below) exemplifies this.

Chocolate Oatmeal Cookies

| | |
|---|---|
| 1/4 cup unsweetened cocoa | 2 cups tabletop aspartame |
| 1/2 cup skim milk | 3 cups oatmeal |
| 1/2 cup diet margarine | 1/4 cup coconut, if desired |
| 1/4 tsp vanilla | 2 tbs chopped nuts, if desired |

Combine skim milk, cocoa, margarine, and vanilla in saucepan. Simmer at low temperature and stir until margarine is completely melted. Remove from heat. Add sweetener and oatmeal (plus coconut and chopped nuts, if desired). Stir until mixed. Using teaspoon, drop onto cookie sheet. Refrigerate for 1 hr. Makes 50–60 cookies. This recipe yields 25–35 cal/cookie. Recipe made with traditional ingredients yields 56–60 cal/cookie.

**B. Dry-mix Products**

*1. General*

This category is ideal for low calorie foods formulated with aspartame. A variety of such products have been formulated:

- Cold beverage mix
- Gelatin dessert mix
- Pudding and pie filling mix
- Hot chocolate powder
- Whipped topping mix
- Blender shake mix
- Salad dressing mix
- Vegetable or chip dip mix
- Blender dessert mix
- Plain yogurt flavoring mix
- Instant presweetened coffee or tea powder
- Chocolate flavoring for milk

*2. Technology*

As with the tabletop sachet product, packaging plays a key role in product performance following normal storage. The proper selection of a moisture barrier (on a technical basis) will more often relate to ingredients other than aspartame, but all must be considered.

Rate of dry mix dissolution is another important area, and Malto-Dextrins[146] have been used to prepare free flowing, readily soluble beverage mixes. To prevent separation of dry-blend beverages, a dried paste of aspartame and calcium biphosphate can be added.[147] Dry-mix products can also be prepared as blends when including calcium carbonate as a dry carbonation source, a food acid, aspartame, and flavorings.[148]

*3. Formulations*

***a. Cold Beverage Mixes***

Beverage mixes are ideally suited to calorie reduction in quality products. Several beverage mix products have been prepared; formulations, preparation ratios, and calorie reduction are shown in Table 10.

***b. Gelatin Dessert Mixes***

In the beverage mixes just discussed, aspartame replaced sweetness. "Body" was not a problem, but could easily have been provided by gums if necessary. In a gelatin dessert, texture and texture stability are integral aspects of product quality. Sugar partially provides the textural properties expected in a gelatin dessert. In formulating a low calorie product, the texture and its changes during storage must be carefully studied in relation to sugar removal and in relation to addition of aspartame or any other new ingredient. A typical formulation is shown in Table 11.

Partial sugar replacements can be easily developed, and preliminary work shows that aspartame improves the flavor of a sucrose product when a portion of the sugar is replaced with sufficient aspartame to create an "isosweet" product.

TABLE 10

**Dry-mix Beverages – Formulation, Dilutions, and Calorie Reduction**

| Ingredients | Composition of dry-mix (%) Pink lemonade | Lemonade | Strawberry | Orange |
|---|---|---|---|---|
| Citric acid | 79.70 | 64.90 | 45.70 | 42.30 |
| Mono calcium phosphate | – | 20.50 | 38.10 | 37.20 |
| Clouding agent | 5.75 | 3.88 | – | – |
| Ascorbic acid | 0.32 | 0.78 | 1.40 | 1.28 |
| Vitamin A | – | 0.07 | 0.13 | 0.13 |
| Flavors | 3.00 | 2.68 | 0.64 | 5.77 |
| Colors | 0.30 | 0.01 | 1.27 | 0.51 |
| Aspartame | 10.90 | 7.20 | 12.70 | 12.80 |
| Total | 100 | 100 | 100 | 100 |
| **Mix in Finished Beverage (%)** | | | | |
| | 0.58 | 0.75 | 0.41 | 0.41 |
| **Calories Compared to a Sugar Counterpart (%)** | | | | |
| | 5.0 | 5.8 | 2.5 | 2.6 |

TABLE 11

**Gelatin Dessert – Formulation, Dilution, and Calorie Reduction**

| Ingredients | Composition of dry-mix (%) |
|---|---|
| Gelatin | 67.9 |
| Acidulant (adipic/fumaric) | 19.5 |
| Sodium citrate | 6.4 |
| Aspartame | 3.8 |
| Flavor | 2.1 |
| Color | 0.3 |
| | 100 |

*Note:* Dry-mix in finished dessert = 2.4%; calories compared to sugar counterpart = 12.4.%.

TABLE 12

**Chocolate Pudding or Pie Filling – Formulation, Dilution, and Calorie Reduction**

| | Composition of dry-mix (%) |
|---|---|
| Starch thickeners | 68.0 |
| Cocoa | 12.9 |
| Alginates | 6.5 |
| Trisodium phosphate | 5.2 |
| Aspartame | 2.6 |
| Sodium dihydrogen phosphate | 2.3 |
| Calcium acetate | 1.9 |
| Sodium chloride | 0.6 |
| Flavor and color | (Minor) |
| | 100 |

*Note:* Dry-mix in finished milk based chocolate pudding = 6.2%; calories compared to sugar counterpart = 43%.

### *c. Pudding and Pie Filling Mixes*

Considerations for gelatin desserts also apply to puddings; however, puddings are usually set with alginates and starches. In gelatin desserts, gelatin is the sole thickener. A typical low calorie pudding and pie filling formulation is shown in Table 12. There is less caloric reduction in this example because of the fixed caloric contribution from milk used in pudding preparation.

### *d. Hot Chocolate Mixes*

Formulations for diet, chocolate-flavored dairy products must be carefully defined. The product can be dissolved either in water or milk, and the formulations and technological challenges of each liquid are quite different. Furthermore, if the product is to be used in milk, dissolution in both hot milk for hot chocolate and cold milk for chocolate milk must be considered. The dissolving of the powder mix is a major consideration and is accomplished by the presence of sugar in a nondiet

TABLE 13

**Hot Chocolate Mix Formulation, Dilution, and Calorie Reduction**

| Ingredients | Composition of powder (%) |
|---|---|
| Nonfat dry milk solids | 75.0 |
| Cocoa (lecithinated) | 22.5 |
| Flavoring (vanilla and enhancer) | 1.5 |
| Aspartame | 1.0 |
| | 100 |

*Note:* Dry-mix in finished drink prepared with water = 8.3%; calories compared to sugar counterpart = 55%.

TABLE 14

**Whipped Topping Mix – Formulation, Dilution, and Calorie Reduction**

| Ingredients | Weight composition of mix (%) |
|---|---|
| Instant dehydrated cream substitute | 67.4 |
| Nonfat dry milk solids | 28.8 |
| Topping flavor (spray-dried) | 2.4 |
| Aspartame | 0.9 |
| Surfactant | 0.4 |
| Vanilla (spray-dried) | 0.1 |
| | 100 |

*Note:* Dry-mix in finished topping prepared with water (8 min in home mixer) = 26%; calories compared to sugar counterpart = 39%.

formulation. In formulations used with water, milk solids serve this purpose. The use of lecithinated cocoa also aids dissolution.

Other factors to be considered are the need to reduce calories contributed by chocolate, as sugar removal alone is insufficient for 50% calorie reduction, and to satisfactorily compensate for flavor loss with vanilla flavorings and other extenders. It may also be necessary to add bodying agents to compensate for the almost 50% reduction in solids. Table 13 shows an unoptimized hot chocolate formula for use in hot water.

### *e. Whipped Topping Mix*

A properly formulated whipped topping must whip, stay whipped, and have the proper taste and mouthfeel. Replacing sugar solids with anything edible will not result in caloric reduction; hence, the challenge is to formulate a product with body but without solids. Table 14 shows a dry-mix whipped topping with 4.7 cal/tbs vs. 12 cal/tbs for a commercial sucrose-sweetened product.

### *f. Blender Shake Mix*

The calorie content of milk shakes ranges from 300 to 700 cal/16-oz serving. This suggests an excellent opportunity to develop a low calorie product. The temperature, processing, and storage of commercially prepared shakes are all compatible with aspartame's properties. The ice cream's frozen properties can be replaced with ice cubes or unsweetened frozen fruit and cold skim milk. Gums can provide body and aeration.

Two basic formulas have been developed which yield a 50% calorie reduction by volume and are percieved as thick with acceptability even 15 min after preparation. The details are shown in Table 15. Note that caloric content of marketplace milk shakes varies substantially, but the caloric reduction is significant even against the lower calorie commercial products.

Three flavored formulations (vanilla, chocolate, and strawberry) were tested on a single stimulus basis in a design which incorporated as controls: (1) a full-calorie, popular, fast service franchise shake and (2) a packaged home-use, calorie-reduced product. Both controls were vanilla flavored. Intention-to-buy judgments were obtained from 654 women concerned with weight loss. They believed all products were low calorie. The results are shown in Table 16.

The respondents preferred the caloric product on an acceptability basis, but the aspartame prototype products received higher ratings than their commercial saccharin counterparts.

### *g. Salad Dressing Mix*

Low calorie salad dressings present a very unlikely opportunity for use of an intense sweetener because most calories in dressings come from salad oil (40 to 70% by weight), and most salad dressings are not sweet. A number of available low calorie dressings contain altered proportions of oil and water. Body is created with surfactants and gums. Coleslaw dressing tastes sweeter than other dressings and need not be high calorie just because of salad oil. A calorie-reduced, aspartame-sweetened formulation is shown in Table 17. Some French dressings are also sweet,

TABLE 15

**Blender Shake Mixes: Formulation, Dilution, and Calorie Reduction**

| | Weight composition of mix (%) | |
|---|---|---|
| | Flavored | With frozen fruit to be added separately |
| Aspartame (3.45% in spray-dried Malto-Dextrin) | 58.8 | 68.6 |
| Modified starches | 26.1 | 19.6 |
| Xanthan gum | 4.2 | 4.9 |
| Microcrystalline cellulose | 4.2 | 4.9 |
| Color (1% solution) | 2.9 | – |
| Trisodium pyrophosphate | 1.7 | 1.7 |
| Carrageenan | 0.8 | – |
| Gum karaya | 0.8 | – |
| Banana flavor | 0.4 | – |
| | 100 | 100 |

Dry-mix in finished shake (%)
16 oz from 11.6 g + (1 cup skim milk + 4 ice cubes) = 2.8%
16 oz from 8.8 g + (1 cup skim milk + 1 cup frozen berries) = 2.1%

| Calories compared to sugar counterpart (for 16 oz.) | |
|---|---|
| Aspartame product, flavored | 131–136 |
| Aspartame product, fruit | 175–210 |
| Chocolate milk shake | 693[a] |
| MacDonalds, by analysis, chocolate, strawberry or vanilla | 320 |
| Hardee's milk shake | 640[b] |

[a]United States Department of Agriculture USDA Home and Garden Bulletin G–153, Calories and weight, USDA pocket guide, Superintendent of Documents, Washington, D.C., 1970.
[b]From How to do anything better guide, *Glamour,* May 1973. Copyright 1973 by Condé Nast Publications Inc. With permission.

TABLE 16

**Purchase Intent for Low Calorie Shakes**

| Flavor types | Judges showing an intention to buy (%)[a] |
|---|---|
| Vanilla | |
| Fast-service franchise product (caloric) | 83 |
| Aspartame-based product (low calorie) | 62 |
| Saccharin-based packaged product (low calorie) | 56 |
| Strawberry | |
| Aspartame-based product (low calorie) | 72 |
| Chocolate | |
| Aspartame-based product (low calorie) | 66 |

[a]Total percent respondents indicating probable or definite intention to buy on a five point balanced hedonic scale.

TABLE 17

**Cole Slaw Dressing Mix Formulation, Dilution and Calorie Reduction**

| Ingredients | Weight composition of mix (%) |
|---|---|
| Pregelatinized starch | 32.7 |
| Aspartame (3.45% spray-dried in Malto-Dextrin) | 23.0 |
| Salt | 19.7 |
| Nonfat dry milk solids, instant | 17.7 |
| Mustard flour | 5.9 |
| Soluble pepper on salt | 1.0 |
| | 100 |

*Note:* Dry-mix in finished dressing (30.5 g mix in 2 oz salad oil plus 2 oz vinegar and 4 oz water) = 13%; Calories compared to commercial sugar counterpart = 54%.

TABLE 18

**French Onion Party Dip Mix – Formulation, Dilution, and Calorie Reduction**

| Ingredients | Weight composition of mix (%) |
|---|---|
| Powdered sour cream | 32.1 |
| Plain chopped onions (dry) | 17.6 |
| Hydrolyzed vegetable proteins | 10.4 |
| Toasted chopped onions (dry) | 7.6 |
| Salt | 7.0 |
| Pregelatinized modified starch | 5.1 |
| Autolyzed yeast extract | 4.9 |
| Gelatin | 3.6 |
| Monosodium glutamate | 2.9 |
| Onion powder | 2.0 |
| Aspartame (3.45% in spray-dried Malto-Dextrin) | 1.3 |
| Disodium phosphate | 1.0 |
| Sodium citrate | 1.0 |
| Titanium dioxide | 1.0 |
| Sodium bicarbonate | 1.0 |
| Beef extract powder | 0.50 |
| Imitation sour cream flavor | 0.25 |
| Ground celery | 0.20 |
| White pepper | 0.20 |
| Locust bean gum | 0.20 |
| Garlic powder | 0.15 |
| Silicon dioxide | 0.05 |
| Partially hydrogenated vegetable oil | 0.05 |
| | 100 |

*Note:* Dry-mix in finished dip prepared with plain low-fat yogurt = 14%; calories compared to commercial counterpart = 50%.

but they are standardized products and require optional ingredient approvals.

### *h. Dip Mixes*

Dip mixes usually contain about 2 cal/g and chips contain 10 to 12 cal each. By eating a calorie-reduced dip with vegetable sticks, a significant reduction in calorie consumption can be achieved. Some very acceptable dry-mix products have been formulated for mixing with yogurt, creamed cottage cheese, or water. Table 18 shows the formulation for a French onion dip prepared in a yogurt base. Note that even though aspartame contributes to calorie reduction, this is not a highly sweet product.

*4. Other Sweetened Dry-mix Products*

In addition to those products detailed, additional opportunities for low calorie products exist. A blender mix based on instant pudding technology can be made by straightforward sugar substitution. Instant coffee or tea can easily be presweetened; however, it may not be easy to select one optimal sweetness level satisfying all consumers. A powdered, sweetened, chocolate flavoring for milk should be possible even though the sugar does more than add sweetness to the conventional products.

Flavoring plain yogurt with a dry mix of flavor, color, and sweetener is feasible. The dry mix could be prepared in portion packs like a table sweetener sachet and stirred into plain yogurt. The caloric addition to plain yogurt is almost negligible, but the importance of the visual absence of fruit pieces must be assessed (see Table 19).

Examples discussed in this section are only indications of the diverse opportunities existing for development of good quality, low calorie, dry-mix foods.

TABLE 19

**Dry-mix Flavoring for Converting Plain Yogurt to Cherry-flavored Yogurt – Formulation, Dilution, and Calorie Reduction**

| Ingredients | Weight composition of mix (%) |
|---|---|
| Malto-Dextrin carrier | 91.6 |
| Aspartame | 5.4 |
| Artificial cherry flavor | 2.4 |
| Colors | 0.6 |
| | 100 |

*Note:* Dry-mix in finished flavored yogurt = 1%; calories compared to sugar counterpart = 50%.

## C. Beverages

*1. Carbonated Beverages*

Beverages, particularly carbonated soft drinks, constitute an enormous market for sugar and provide the largest application of low calorie sweeteners. Table 2 shows 18.8 lb of sugar per capita per year consumed in beverages annually – at least 250 bottles of soda per person per year. The use of aspartame in low calorie beverages can be achieved by matching commercial practices such as pH, processing temperatures, distribution, use of concentrates, and storage conditions with product quality and lifetime requirements. Calories can be reduced by 90% or more.

***a. Dry Sweetener Pre-mix for Bottlers***

The concept was developed to combine dry-mix standard ingredients in a blend adapted for use by soft drink bottlers. Table 20 shows the blend's composition and the drink prepared from it. A "syrup" is prepared using a flavored beverage base, water, and the blend. One part of that syrup is blended with five parts carbonated water.

***b. Relationship of Aspartame Stability and pH in Soft Drinks***

Early pH-stability studies carried out in buffers and shown in Figures 10 to 13 raised doubts about aspartame application in soft drinks. To determine if stability in soft drinks was technically feasible, or related to formulation ingredients, a study was carried out by G. D. Searle & Co., Skokie, Ill., and several commercial bottlers. The bottlers provided flavor concentrates and sometimes acidulants and preservatives along with preparation and carbonation instructions for the products. Various soft drinks were prepared, and analytical procedures were developed prior to the study's initiation.

TABLE 20

**Bottlers' Blend for Carbonated Soft Drinks**

| Ingredient | Purpose | Blend (%) | Finished beverage (%) |
|---|---|---|---|
| Aspartame | Sweetening | 44 | 0.053 |
| Sodium benzoate | Preservation | 41 | 0.050 |
| Sodium carboxymethyl cellulose | Bodying | 15 | 0.018 |
| | | 100 | |

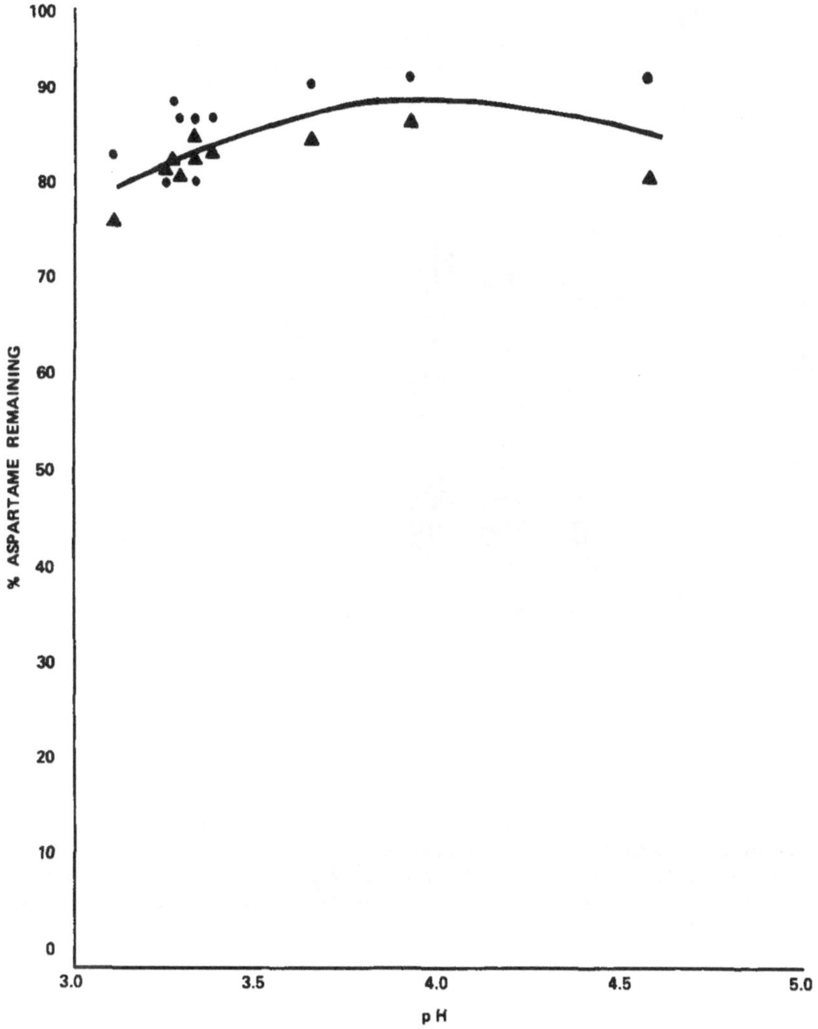


FIGURE 15. Effect of pH on aspartame stability in various soft drinks. Each point represents a commercial sample prepared to manufacturers' specifications and aspartame (circles) or radiolabeled aspartame (triangles) was added at approximately 0.05%. Recovery data are shown after samples were stored 60 days at room temperature.

Aspartame was used to sweeten one set of samples bottled normally and stored at room temperature. Radio-labeled aspartame was added to some of the unsweetened soft drinks sealed in vials and stored at room temperature. Flavors included orange, root beer, cola, lemon-lime, and ginger ale. Aspartame recoveries from this experiment are shown in Figure 15. Though the wet chemical and radiolabeled analyses show minor variance, apparently formulation differences (aside from pH) have little effect. On the other hand, an attempt to formulate drinks at pH levels above 3.5 could substantially improve the possible success of aspartame use in soft drinks. Though a controlled experiment was not conducted, acidulants do not seem to show significant stability differences.

### *c. Effect of Aspartame Loss on Soft Drink Acceptability*

Because conversion products can be predicted when aspartame is formulated in soft drinks and stored under normal conditions, a study was conducted to determine the effect of this loss of sweetness. Three sets of drinks were prepared (cola, lemon-lime, and root beer) from commercial concentrates, and a saccharin commercial control was included in the study. The aspartame products were equal in sweetness to the saccharin products,

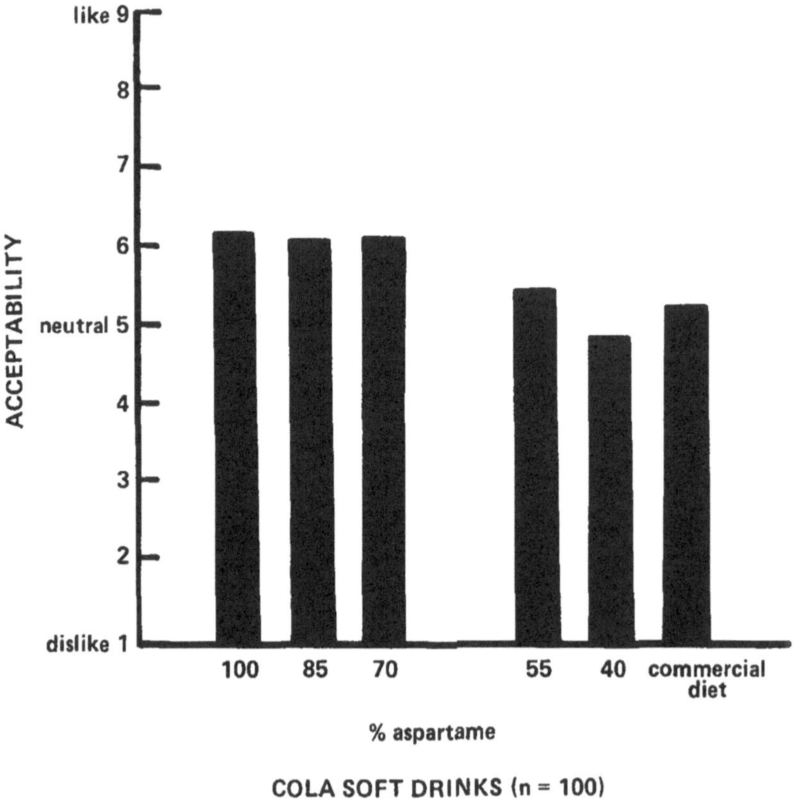


FIGURE 16. Acceptability of various unstored aspartame-sweetened cola drinks vs. a commercial saccharin control. Drinks differ only in sweetness source. 100 = preparation to be isosweet with saccharin control; 85 = 85% of the aspartame concentration in 100, etc. (pH = 3.0 to 3.1).

and also intentionally reduced in sweetness by 15% decrements in aspartame down to a level of 40% of that used for saccharin isosweetness (100, 85, 70, 55, and 40%).

These fresh samples were hedonically tested for acceptability. Each panelist judged only one sample. Each data point represents 100 judgments. The cola beverage data in Figure 16 establish some important quality differences between aspartame and saccharin drinks. Only after about half of the aspartame is removed from an isosweet drink is its acceptability as low as that of the commercial saccharin product. A drink with 70% aspartame for isosweetness is statistically preferred to the commercial saccharin product. Results for other flavors are similar. Because flavors can deteriorate with time, it is also important to determine the acceptability of stored drinks.

*d. Detailed Study of a Cola Drink*

Because more than half of all soft drinks sold are cola flavored, it is the preferred flavor for study. Most sugar colas have a pH in the range of 2.5 to 2.7. A commercial diet cola formulation with pH 3.1 was selected for study. In this study, three parameters were measured: (1) remaining aspartame, (2) perceived sweetness, and (3) overall acceptability. The sensory evaluations were run at the laboratories of the flavor base manufacturer, and analytical determinations were run at G. D. Searle & Co. Each sensory judgment involved 12 to 13 individuals from a trained group of 16. The results are shown in Figure 17. As anticipated, there was a loss of aspartame, but after 6 months' storage at room temperature, two thirds aspartame remained. The sweetness intensity determinations, shown in the center portion of Figure 17, reflect the aspartame lost, thus indicating a gradual but

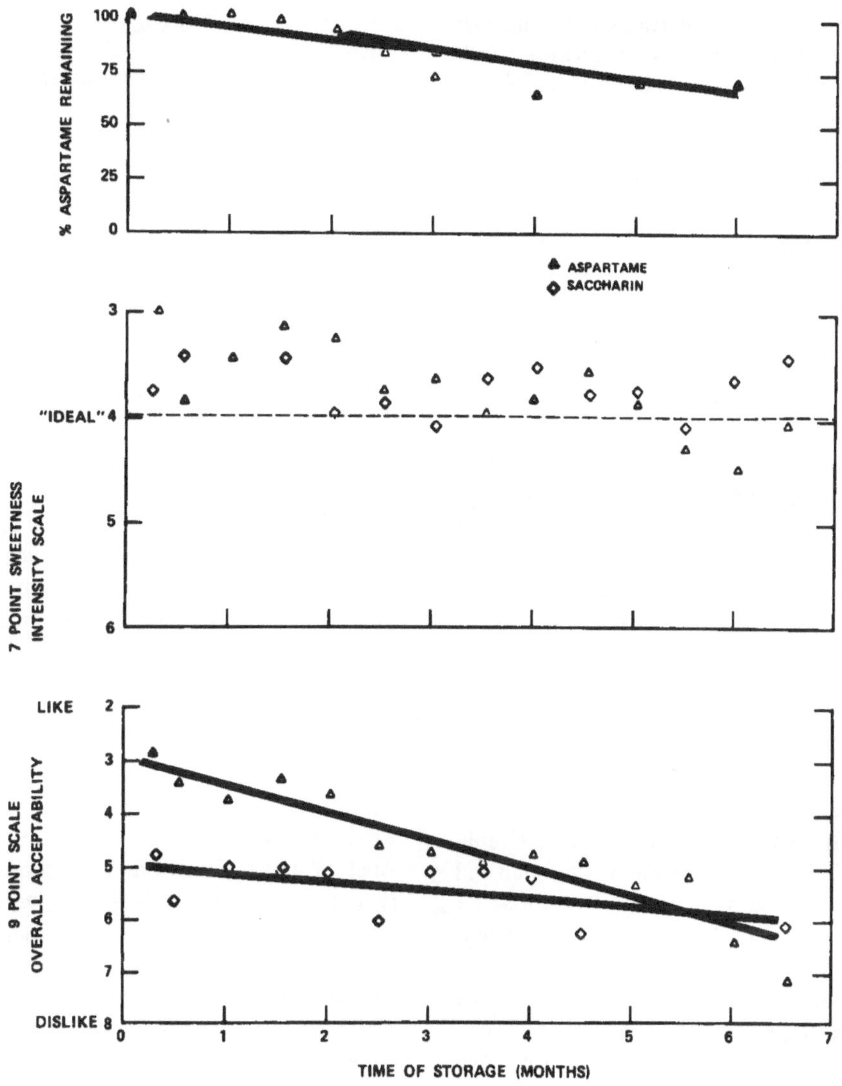


FIGURE 17. Storage stability study of an aspartame-sweetened cola drink (pH = 3.1) Top: chemical stability of aspartame; middle: stability of sweetness intensity; bottom: stability of acceptability. Samples were stored at room temperature.

very slight reduction in perceived sweetness from nearly "slightly too sweet" just after bottling to "ideal" or "sweetness just right" after 5 months. Sweetness of the commercial saccharin control product was stable over 6 months' storage at just sweeter than "ideal."

The aspartame loss shown in the top graph of Figure 17 is partly perceived as a sweetness loss like one shown in the middle graph, but it does demonstrate an acceptability loss as shown in the bottom graph. The acceptability data demonstrate that loss of acceptability may be due partly to flavor loss since the acceptability of a saccharin-sweetened beverage decreases with storage life. Initially, the aspartame product was clearly preferred to its saccharin counterpart, and only after 5 months had the aspartame product dropped to the acceptability of the commercial saccharin counterpart. To put this time period into perspective, it is important to realize that the vast majority of bottled or canned soft drinks are consumed within 2 to 3 mo of manufacture. Cans presumedly circulate slightly longer than bottles.

In this study, a third product was tested and sweetened to the same total sweetness level with a mixture of aspartame and saccharin proportioned for equal sweetness contributions. The analysis results and sensory evaluation were consistent with the pure aspartame system data: percent conversion of aspartame with time was similar to conversion in samples sweetened only with aspartame, and acceptability of the blend was between that of soft drinks sweetened with aspartame only and saccharin only. Specifications for the three drinks are shown in Table 21.

TABLE 21

**Aspartame-sweetened Carbonated Soft Drinks**

| Aspartame-sweetened soft drink isosweet to commercial diet product | pH | Aspartame ppm |
|---|---|---|
| 100% sweetness from aspartame | 3.1 | 528 |
| 50% sweetness from aspartame (107 ppm saccharin) | 3.0 | 176 |
| 100% Saccharin control | 2.9 | – |

*Note:* All three products contained the same flavoring system, carbonation, benzoate, etc.

### *e. Acceptability and Sweetness of Cola Drinks Sweetened with Combinations of Aspartame, Saccharin, and Cyclamate*

This study provides a mechanism for predicting sweetness and quality of various binary or ternary mixtures and their associated costs resulting from ingredient costs and synergy savings. A commercial cola concentrate was used to prepare about 50 drinks which varied in source and intensity of sweetness. Magnitude estimation,[149] judgments of sweetness, quality, degree of popularity, and acceptable purchase price were obtained from 35 women panelists who tasted samples five at a time. Each set contained a control (0.2 g of a saccharin-cyclamate blend 58.5:41.5, per 10 fluid ounces). During the entire test, each evaluation was replicated three times, though samples were coded and a randomized presentation used. In addition, six unidentified commercial controls were tested: Pepsi Cola®, Diet Pepsi®, Royal Crown Cola®, Diet Rite Cola®, Coca Cola®, and Tab®. A number of developed equations related (1) sweetness to acceptability, (2) sweetness to sweetener concentrations, (3) acceptability to sweetener concentrations, (4) off-after taste to sweetener concentrations, and (5) estimated purchase price to acceptability.

The interesting relationship of sweetness to acceptability is shown in Figure 18. It establishes a sweetness profile indicating a maximum acceptability at a fixed sweetness, after which acceptability drops. Furthermore, it is clear that all six commercial drinks, whether sweetened with sugar or saccharin, have almost identical sweetness intensity. It is surprising that the sweetness of commercial products is substantially below maximum preference. Possibly the higher sweetness, though more acceptable in these test conditions, results in faster satiety. Furthermore, the diet products sweetened with saccharin would be too bitter at sweeter levels.

### *f. High Temperature Storage*

Some additional storage studies were conducted at 90°F with an aspartame-sweetened commercial soft drink acidified with phosphoric acid, but prepared at its normal pH of 3.3, and one prepared at a higher pH of 3.9. The manufacturer stored and periodically analyzed products for taste stability vs. the stability of the commercial saccharin product at pH 3.3. The manufacturer's test was triangular for difference, followed by a preference determination; preference data were deemed valid only when differences could be detected. Analytical determinations were conducted by G. D. Searle & Co. The data in Table 22 clearly show that raising the pH of the product achieves increased stability. However, for this particular formulation, it also indicates an organoleptic preference for tartness at the lower pH. When comparing experimental products to the commercial saccharin control, experienced tasters (6 or more from a group of 12) demonstrated a preference for the aspartame product at pH 3.3, even when one third of the aspartame had been converted. At pH 3.9 and a 20% loss of sweetener, the preference trend for aspartame disappeared.

### *2. Refrigerated Still Beverages*

Refrigerated still beverages are a smaller, yet significant market for sugar replacement. These products are usually sold in 1- or ½-gal plastic bottles in a nonpremium market. Calories could easily be reduced by 90% or more, but total caloric content is dependent on the proportion of flavors derived from fruit juices or from artificial flavorings.

A number of products have been prepared and evaluated for remaining aspartame, sweetness, and

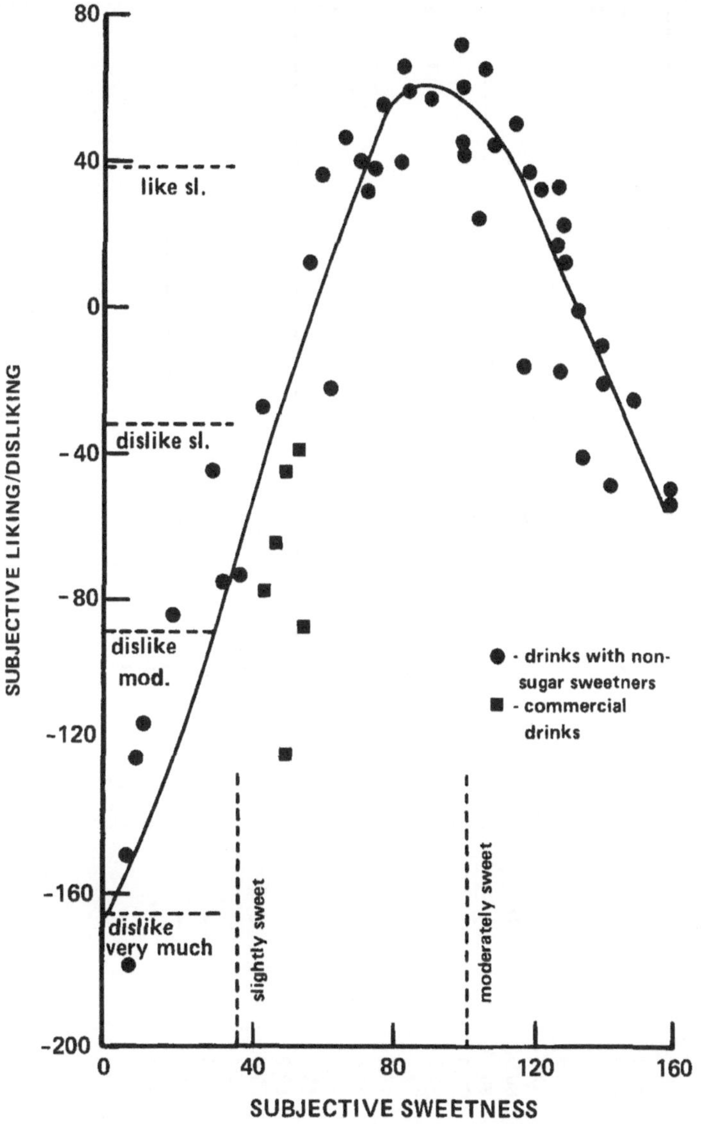


FIGURE 18. Acceptability of a cola beverage (sweetened to various levels with mixtures of aspartame, cyclamate, and saccharin) vs. perceived sweetness and compared to commercial drinks sweetened with saccharin or sugar.

acceptability. The data for an aspartame-sweetened iced-tea product having a pH of 3.6 to 3.7 are shown in Figure 19 and are representative of other flavored formulations: orange, lemon, grape, and fruit punch.

Over a period of 12 weeks of refrigerated storage (more than the expected shelf life for this product), the aspartame tea was stable. Sweetness decreased and the acceptance, though stable, indicated a slight loss. Throughout the taste evaluation, acceptability was near "like slightly." If perceived sweetness loss is due to flavor deterioration, then aspartame is more than sufficiently stable at this pH to match current manufacturing practices. The formulation for the tea-flavored beverage is shown in Table 23.

*3. Liquid Drink Concentrates (1 + 5)**

Three flavors of concentrates were prepared for use as fountain syrup or in still drink dispensers.

*1 part mix, 5 parts water.

TABLE 22

**Aspartame Stability in Soft Drinks at 90°F**

| | pH 3.3 | | pH 3.9 | |
|---|---|---|---|---|
| Storage (weeks) | Aspartame remaining (%) | Taste vs. saccharin[a] | Aspartame remaining (%) | Taste vs. saccharin[a] |
| 0 | – | A | – | A |
| 1 | 94.9 | – | 96.2 | – |
| 2 | 89.8 | A | 92.9 | A |
| 5 | 79.0 | – | 86.7 | – |
| 6 | 74.2 | A | 82.3 | – |
| 7 | 73.7 | – | 81.6 | – |
| 8 | 71.2 | – | 80.9 | ND |
| 9 | 68.5 | – | 80.1 | – |
| 10 | 64.2 | – | 79.4 | – |
| 11 | 63.2 | – | 76.1 | – |
| 12 | – | A | 71.2 | ND |
| 13 | – | – | 73.6 | – |
| 14 | – | – | 73.4 | – |
| 15 | – | – | 67.5 | – |
| 16 | – | – | 69.7 | – |
| 17 | – | – | 67.0 | – |
| 18 | – | – | 67.4 | – |
| 19 | – | – | 64.1 | – |
| 20 | – | – | 63.7 | – |
| 21 | – | – | 61.5 | – |
| 23 | – | – | 60.4 | – |

*Note:* A, aspartame preference trend; ND, no difference.

[a] Control at pH 3.3 was a commercial phosphoric acid drink.

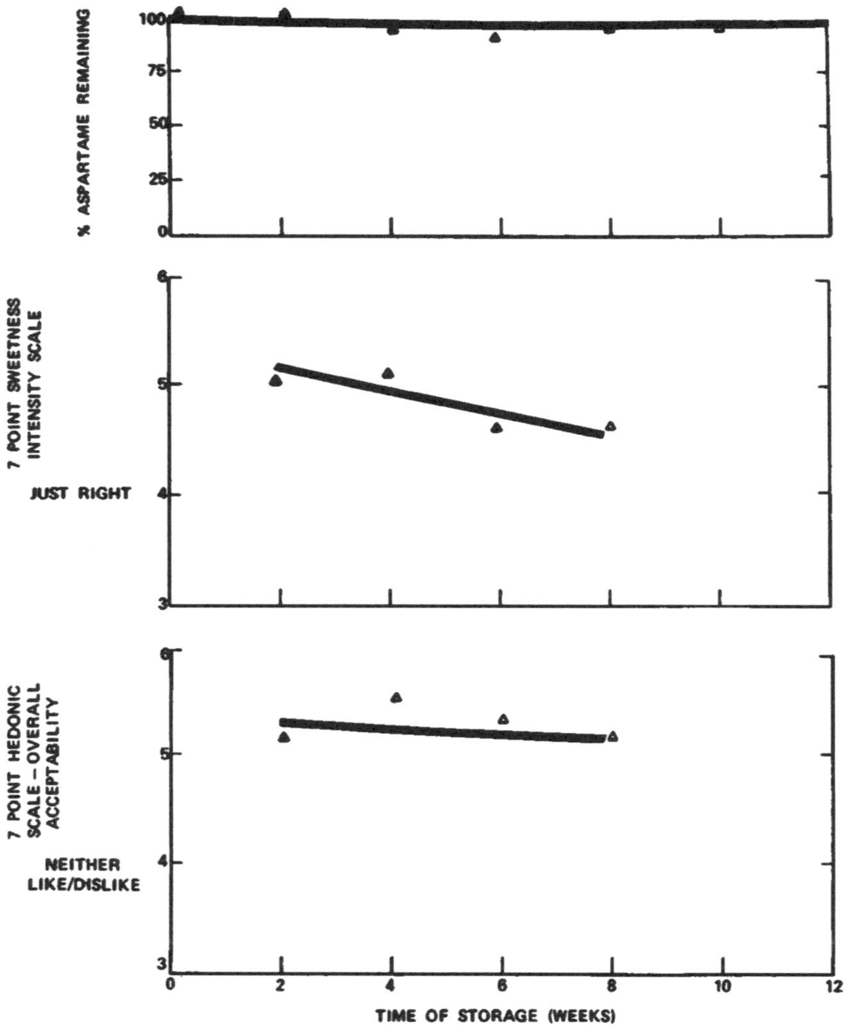


FIGURE 19. Storage stability study of an aspartame-sweetened, iced-tea-flavored, refrigerated still beverage (pH = 3.6 to 3.7). Top: chemical stability of aspartame; middle: stability of sweetness intensity; bottom: stability of acceptability.

The concentrate is mixed with five parts water by volume for consumption. The composition of a pink lemonade concentrate is shown in Table 24. Lemon-lime and root beer formulas have also been developed.

*4. Pasteurized Fruit Juice Drinks*

An orange juice drink, containing 10% juice at pH 3.6, was sweetened with 0.04% aspartame and used for a pasteurization test in the laboratories of a commercial juice firm. The pilot conditions estimated to be more severe than those in a large-scale, high-temperature, short-time unit were to bring 3 qt of the product from room temperature to 190°F in 6 min in an electric steam kettle. The drink was then bottled and cooled in a sink with 55°F water for about 3 hr until room temperature was reached. The aspartame in this drink was found to be analytically stable to the described processing (see Table 25).

TABLE 23

**Composition of Tea-flavored Refrigerated Still Beverage**

| Ingredients | Composition in finished beverage (%) |
|---|---|
| Instant tea | 0.202 |
| Lemon crystals | 0.178 |
| Citric acid, anhydrous | 0.143 |
| Sodium citrate, hydrate | 0.071 |
| Sodium chloride | 0.071 |
| Aspartame | 0.045 |
| Sodium benzoate | 0.024 |
| Color | 0.0003 |
| Total solids | 0.734 |

TABLE 24

**Composition of a Pink Lemonade Concentrate[a]**

| Ingredients | Composition in concentrate (%) |
|---|---|
| Citric acid | 2.54 |
| Lemon cloud emulsion | 0.76 |
| Aspartame | 0.37 |
| Sodium benzoate | 0.25 |
| Colors | 0.10 |
| Total ingredients | 4.02 |

[a] 1 part concentrate + 5 parts water.

TABLE 25

**Processing Stability of Aspartame in a Processed Orange Drink (pH 3.6)**

| Sample | Aspartame (%) |
|---|---|
| Theoretical addition | 0.040 |
| Unpasteurized orange drink | 0.0402 ± 0.001 |
| Pasteurized orange drink | 0.0415 ± 0.001 |
| Pasteurized orange drink (adjusted for maximum moisture loss in processing –5.2%) | 0.0394 ± 0.001 |

**D. Bakery, Cereal, and Allied Products**

*1. Baked Goods*

**General** – Replacing sugar in baked goods involves the replacement of sweetness, bulk, browning, yeast, food, and humectancy properties. Even if a suitable bulking material were available, the baking process requires prolonged exposure to moist heat at pH levels generally well above 4.5, and the use of aspartame in baking will require substantial creativity. Analytical recovery of aspartame or of its conversion products from dough systems has been difficult and has restricted quantitative research. Two organoleptic studies have shown substantial, but not total, sweetness loss in baked goods sweetened with aspartame. The systems were bread and a cinnamon-apple pie. In both studies products were prepared without any sweetener, and with sucrose and with aspartame at 1/180 the sucrose weight. Organoleptic results established that the sucrose product was sweeter than the aspartame product, indicating some conversion. The aspartame product was sweeter than no sweetener, indicating residual sweetness.

**Aroma** – Aspartame has been used in conventionally heated bakery goods at 0.005 to 0.5% of the starting starch for the purpose of aroma improvement.[150] This was not an expected use.

**Dough conditioning** – In a yeast-raised white bread formula, aspartame performs much like a dough conditioner. Presumedly, aspartame's primary function is to sweeten, and its function as dough conditioner is an additional benefit.

**No-bake products** – Some nontraditional approaches have been used to formulate unbaked, sucrose-free products as substitutes for baked goods. One approach involves mixing microcrystalline cellulose, starch, lipids, and gums with

TABLE 26

**Aspartame-sweetened Allied Products**

| Product | Storage temperature (°F) | pH | Moisture (%) | Aspartame formula (cal/g) | Sugar formula (cal/g) | Comments |
|---|---|---|---|---|---|---|
| Coating, powdered, doughnut | 72 | 5.5 | 3.3 | 3.8 | 4.0 | Aspartame sample had superior performance |
| Filling, cream, marshmallow | 40 | 5.4 | 26 | 3.8 | 3.8 | Contains sorbitol |
| Filling, pie, apple | 40 | 3.3 | 85 | 0.8 | 1.3 | Aspartame sample lacks glassy surface |
| Filling, Pie, lemon | 40 | 3.3 | 91 | 0.5 | 1.0 | – |
| Frosting, butter cream | 72 | 6.0 | 19 | 3.0 | 4.0 | – |
| Glaze, confectioners | 72 | 5.0 | 1.6 | 5.3 | 5.5 | – |
| Glaze, doughnut | 72 | 5.6 | 11 | 2.5 | 2.7 | Lactose base |
| Jelly, bakers | 40 | 4.1 | 78 | 1.8 | 1.9 | – |
| Topping, streussel | 72 | 6.0 | 3.0 | 4.1 | 4.5 | – |

flavors and aspartame in a milk base; leavening chemically; and then setting by refrigeration. A lemon bar is an example of an unbaked, aspartame-sweetened product. Another formulation method involves preparing a binder of water, lipids, starch and cellulose, surfactants, and aspartame, whipping the cooled binder before adding bread crumbs (one third by weight), and then refrigerating.

*2. Breakfast Cereal*

Most breakfast cereals are consumed by children who do not need caloric restriction. Sweetening cereal with a high potency sweetener is not done to reduce calories. Replacing the sugar in the cereal should improve its nutritional content by removing the nutrient dilution effects of sugar. When sugar is removed from cereal, cereal grain replaces its bulk if aspartame can be used for sweetening.

Aspartame can be used to sweeten breakfast cereals by including it in the cereal mix just prior to forming, cutting, and drying. It can be applied as a coating to cereals processed under more severe conditions. Removing sugar from the cereal mix could result in loss of tenderness in the product; removing sugar from cereal coating removes the sparkle, thus altering its appearance, and removing sugar from the formula may raise the cost of goods sold. Apparently, these problems can be solved by total and partial sugar replacement.

"Sticking" aspartame to the cereal surface is challenging and a number of patents have been issued describing different methods. One method, specifically related to cereals, involves the application of a uniform dispersion of aspartame in a starch hydrolyzate,[151] while another method applies a foam formed from an aspartame-saturated solution to the surface of the unsweetened base (0.01 to 5%), and removes water at no more than 160°F.[152]

*3. Allied Bakery Products – Fillings, Glazes, and Toppings*

The baking industry uses a variety of sweet accessories that can contribute substantially to sucrose content and total calories. Though only preliminary work has been completed, a number of aspartame-sweetened prototypes have shown promise. Table 26 lists aspartame-sweetened allied products along with some of their properties. Though an intensive effort was not made to reduce calories, only the pie fillings were significantly reduced. All formulations are sucrose-free.

The aspartame-sweetened powdered doughnut coating has some distinct advantages over sugar counterparts. This coating has greater stability under normal moist storage conditions. On storage, the sucrose product loses its white-powdered appearance and becomes sticky, but the aspartame coating remains a white-powdered sweet coating. Table 27 shows the coating composition.

TABLE 27

**Aspartame-based Powdered Doughnut Coating**

| Ingredients | Composition (%) |
|---|---|
| Corn starch | 95.6 |
| Partially hydrogenated vegetable oil | 3.8 |
| Aspartame | 0.6 |
| | 100 |

TABLE 28

**Calorie-reduced Cheesecake Composition and Calories**

| Ingredients | Composition (%) |
|---|---|
| Filling | |
| Milk | 62.9 |
| Dry-mix portion | |
| Bakers cheese, dehydrated | 10.7 |
| Fat, spray-dried with surfactants | 5.33 |
| Alginate gum | 1.78 |
| Tetrasodium pyrophosphate, anhydrous | 0.355 |
| Calcium gluconate | 0.355 |
| Cheesecake flavor, spray-dried | 0.231 |
| Methyl ethyl cellulose | 0.190 |
| Adipic acid | 0.178 |
| Lemon flavor, spray-dried | 0.088 |
| Aspartame | 0.057 |
| Vanilla | 0.027 |
| Colors | 0.002 |
| Crust | |
| Margarine | 5.33 |
| Dry-mix portion | |
| Graham flour | 12.4 |
| Cinnamon | 0.071 |
| Aspartame | 0.027 |
| Caramel color | 0.009 |
| Vanilla flavor | 0.007 |
| | 100 |

*Note:* Calories/70-g serving; aspartame formula, 150 cal; sugar, cream cheese, sour cream product, over 500 cal.

### *4. Unbaked Cheesecake*

A calorie-reduced indulgent food is open to question, but may fulfill a dieter's need to be human. Several of the nationwide weight loss programs devise menus that include small portions of the dieter's favorite "goodies."

Cheesecake, one of the richest and calorically dense desserts, is a challenging target for caloric reduction. A 70% reduction in calories is still a very rich dessert, but the formula shown in Table 28 exemplifies successful calorie reduction. Even the low calorie cheesecake is sinfully delicious.

## E. Confectionary Products

### *1. General Products*

Since confections are commonly referred to as sweets, one would assume aspartame sweetener to be an ideal ingredient, but confections have such a high proportion of sugar that total sugar replacement in many products is unlikely. Furthermore, candy technology is the art of controlling the physical properties of sucrose and other sugars; most of that art is related to total control of temperature, moisture, and aeration at various processing stages. Since most traditional candy-making involves sugar fusion destructive to aspartame, other nontraditional methods must be used, especially if replacement of most or all of the sugar is desirable.

Economy, sucrose reduction, and caloric reduction are the main reasons for using aspartame in candies. The latter seems unlikely without a suitable bulking agent, but sorbitol and mannitol have been used in candies for many years.

The term "candies" is so broad that it would be worthwhile to classify types. The U.S. Department of Commerce[153] has used these headings: solid chocolate, panned chocolate and/or confectioners' exterior, panned chocolate and/or confectioners' interior, chocolate coating, hard candy, caramel, marshmallow, nougat, creams, fudge, jellies, licorice, bar goods, 5 and 10¢ specialties, packaged goods, bulk goods, and penny goods. These classifications specifically exclude cough drops, chewing gum, roasted peanuts, and unsweetened popcorn.

### *2. Specific Products*

**Overview** – A wide variety of products have been developed incorporating aspartame. Only a few can be formulated with complete sucrose replacement and significant calorie reduction. On the other hand, many excellent products can be prepared if bulking materials are approved and available for them. Some applications of aspartame in confections represent a means of increasing sweetness and therefore palatability only. Some "nutritional candies" must have sugar removed to add protein and other nutrients, and thus more sweetness is required for product acceptance. In several products, flavor potentiation is reported but not conclusively shown.

**Chewing gum** – A great deal of interest has been shown in chewing gum that is normally two thirds sucrose by weight. Calorie-reduced products can be made having extended flavor perception as discussed earlier (Section VII.E.).[117]

TABLE 29

**Aspartame-based Chocolate Coating**

| Ingredients | Composition (%) Commercial control | Aspartame product |
|---|---|---|
| Coating fat | 32.8 | 41.6 |
| Cocoa powder (11% fat) | 16.7 | 18.9 |
| Sucrose, confectioners | 46.4 | 13.9 |
| Dextrin (low dextrose equivalent) | – | 13.9 |
| Nonfat dry milk solids | 4.0 | 10.9 |
| Aspartame | – | 0.7 |
| Salt | – | 0.08 |
| Butter flavor | – | 0.10 |
| Vanilla flavor | – | 0.02 |
| | 100 | 100 |
| Cal/g | 5.3 | 5.6 |

**Chocolate products** – Chocolate products can be made by replacing a major part of the sugar with aspartame. Though the products have good acceptance, they are richer in taste and are more expensive than their commercial counterparts. The sugar must be replaced physically with ingredients such as coating fat, cocoa powder, and dextrins. Table 29 shows a formula for an aspartame-sweetened chocolate coating. This same coating could be calorically diluted on a volumetric basis by preparing a "crunch" bar with a puffed cereal. Satisfactory coatings have also been prepared using 13.6% sorbitol or 25.3% of an unapproved bulking material.

**Marshmallow** – A satisfactory formula for marshmallows was developed by G. D. Searle & Co., Skokie, Illinois, which is isocaloric with traditional sugar products. Half the sweetness was from aspartame. Using sorbitol and total reformulation, a 20% calorie reduction can be achieved, and with a suitable noncaloric bulking agent, a 36% calorie reduction can be achieved. The formulas are shown in Table 30.

**Tootsie Roll®-type products** – A quite acceptable product can be made if both sorbitol and sugar are incorporated to yield 3.3 cal/g. This is estimated to be about a 10% reduction from the traditional "tootsie"-type product. The formula is shown in Table 31.

**Chewable aspartame-sweetened lozenges** – Aspartame should be well suited to dry tablet candies. A chewable lozenge similar to an after-dinner mint has been prepared. Calorie content was not reduced more than 5%. The formulation is shown in Table 32.

**Aspartame-sweetened lemon drop candy** – In preparation of such a product, sucrose is replaced by sweetening corn syrup. A satisfactory product was prepared even though the aspartame was added as the corn syrup cooled to 240 to 250°F on a cooling table. This technique resulted in a one third loss of aspartame, but the perceived sweetness stabilized thereafter for more than 3 months. The formulation is shown in Table 33.

## F. Canned, Bottled, and Frozen Foods, Jams, Jellies, and Preserves

### *1. Canned Fruits*

Aspartame-sweetened canned fruit may satisfy a specific need for fruit in a dieter's menu, and offers improved taste over saccharin-sweetened canned fruit.

Because of the heat normally associated with retorting, it was surprising to find that aspartame-sweetened, canned fruit cocktail showed optimistic early results in a controlled canning experiment. As with the soft drink study, the project was done in cooperation with a commercial manufacturer; G. D. Searle & Co. carried out the analytical portion of the work. The product preparation and sensory evaluation of sweetness and acceptability were done by the manufacturer whose employees were experienced in tasting canned fruit. The composition of the aspartame-sweetened fruit cocktail is shown in Table 34 with processing data that indicate the amount of heat exposure aspartame received.

The control samples in this test included fruit cocktail in heavy syrup, light syrup, and saccharin solution. Figure 20 indicates that processing caused

TABLE 30

**Aspartame-based Marshmallows**

| Ingredient | Composition (%) | | |
|---|---|---|---|
| | Sugar product | Aspartame | Aspartame with sorbitol |
| Corn syrup, 62 dextrose equivalent | 30.4 | 67.4 | 38.9 |
| Water | 23.5 | 15.7 | 18.7 |
| Sugar | 40.6 | 9.0 | – |
| Sugar, powdered | – | 4.5[a] | – |
| Gelatin, 225 bloom | 1.8 | 2.0 | 2.3 |
| Color in solution, Yellow No. 6 | – | 0.56 | – |
| Aspartame | – | 0.34 | 0.40 |
| Flavor, orange | – | 0.22 | – |
| Citric acid, 45% solution | – | 0.22 | – |
| Invert sugar syrup | 3.7 | – | – |
| Flavor, imitation vanilla | 0.007 | – | – |
| Sorbitol, crystalline | – | – | 27.5 |
| Sorbitol, 70% solution | – | – | 12.0 |
| Flavor, artificial peppermint | – | – | 0.10 |
| Color, green | – | – | 0.01 |
| | 100 | 100 | 100 |
| Cal/g | 2.8 | 2.8 | 2.2 |

[a]Could be eliminated from formula.

TABLE 31

**Aspartame-Containing TOOTSIE Roll®-type Product**

| Ingredients | Composition (%) |
|---|---|
| Sugar, confectioners | 54.8 |
| Sorbitol, crystalline | 9.5 |
| Cocoa, 11% fat | 9.1 |
| Starch, instant | 6.6 |
| Water | 6.2 |
| Protein concentrate, dairy | 5.5 |
| Sorbitol, 70% solution | 3.7 |
| Glycerine | 3.7 |
| Aspartame | 0.58 |
| Flavor, butter | 0.15 |
| Flavor, artificial vanilla | 0.11 |
| Gelatin, low bloom | 0.07 |
| Salt | 0.04 |
| | 100 |

about a 20% conversion of aspartame, but that it is quite stable thereafter. Later data indicate that after 45 weeks, approximately half of the original aspartame remains, and at 1 year, one third remains. The sweetness and acceptability of the products over the 6-month period seem quite stable. Even at 1 year, the acceptability of the aspartame product was close to that of the heavy syrup control. The saccharin commercial product consistently received lower ratings. One factor confounding the data somewhat is that the samples are not isosweet – aspartame being sweeter. Realistically, however, aspartame can be sweeter than saccharin because it does not add bitterness.

Results with water-packed purple plums at pH 3.6 are similar to those of the fruit cocktail experiment. A number of factors contribute to the successful use of aspartame in canned goods: pH, processing, required storage life, and sweetness from the fruit sugars. Low-acid products not only provide a more destructive environment for aspartame, but also might require more extensive heat processing to assure safety. By removing only the sugar, water activity increases, and more species of organisms are capable of growth. Calories in canned fruit sweetened with aspartame should be about the same as present diet-packed fruit or about 50% of the calories in a heavy syrup fruit pack. Overall, it appears that quality products can be prepared and that they retain their acceptance throughout reasonable storage, despite substantial analytical losses of the aspartame sweetener.

*2. Bottled Pickles*

A surprising amount of sugar is used in sweet pickles. Dill pickles have 0.1 cal/g and sweet

TABLE 32

**Aspartame-based Chewable Lozenge**

| Ingredient | Composition (%) | |
|---|---|---|
| | Sugar formula | Aspartame formula |
| Sugar, confectioners | 89.9 | 58.2 |
| Dextrin, low dextrose equivalent | – | 11.7 |
| Starch, instant | – | 11.7 |
| Water | 7.2 | 9.0 |
| Sorbitol, crystalline | – | 5.0 |
| Glycerine | 2.8 | 3.5 |
| Aspartame | – | 0.42 |
| Flavor, lemon | – | 0.18 |
| Citric acid, 45% solution | – | 0.14 |
| Gelatin, low bloom | 0.07 | 0.049 |
| Color solution, yellow No. 5 | – | 0.001 |
| | 100 | 100 |

TABLE 33

**Aspartame-Sweetened Lemon Drop**

| Ingredients | Composition (%) | |
|---|---|---|
| | Sugar control formula | Aspartame formula |
| Sugar | 39.6 | |
| Corn syrup | 48.4 | 98.9 |
| Water | 11.5 | 0.49 |
| Citric acid, hydrous | 0.39 | 0.40 |
| Lemon oil | 0.088 | 0.09 |
| Citric acid, anhydrous | 0.005 | 0.005 |
| Aspartame | – | 0.099 |
| | 100 | 100 |

TABLE 34

**Aspartame-sweetened Canned Fruit Cocktail**

| Ingredients | Composition (%) |
|---|---|
| Fruit mixture | 70.05 |
| Water | 25.75 |
| Pineapple juice | 4.12 |
| Aspartame | 0.07 |
| | 100 |

*Note:* Aspartame solution is added to fruit which was raised to 170°F at pH 4.0 over a 35-min period; the mixture was further processed at 232°F for 15 min in a rotary cooker.

pickles have 1.5 cal/g. As with many other systems, it is not sufficient to replace only the sweetness. Because pickles are acidic (pH 3.2), the increase in water activity is not serious, but the loss of soluble solids is quite important to sweet pickle crispness. Mixed sweetener systems may be necessary to solve sweetness and texture problems unless approved bulking materials become available.

Round-cut pickle samples were prepared by a commercial manufacturer using about 0.08% aspartame in the brine which was hot-filled over the pickles. In less than 9 weeks (the first analysis following zero time), the aspartame had equilibrated in the pickles such that the percent weight in the brine and in the drained pickles was about the same. Samples were stored at room temperature, and aspartame analyses were run by G. D. Searle & Co., while sensory data were developed by the manufacturer's panel. A commercial sugar pack was used as a control.

Over a 5-month period, perceived sweetness was apparently stable despite an analytical loss of nearly 50%. Product crispness had definitely dropped by the 4th month and with it, overall acceptability. The manufacturer is confident that the texture problem can be solved.

*3. Frozen Orange Drink Concentrate (1 + 3)**

Frozen fruit drink concentrates are well suited to aspartame sweetening. The pH is compatible; they are stored frozen, and the juices or drinks are rarely exposed to any heat processing prior to fill.

*1 part concentrate plus 3 parts water.

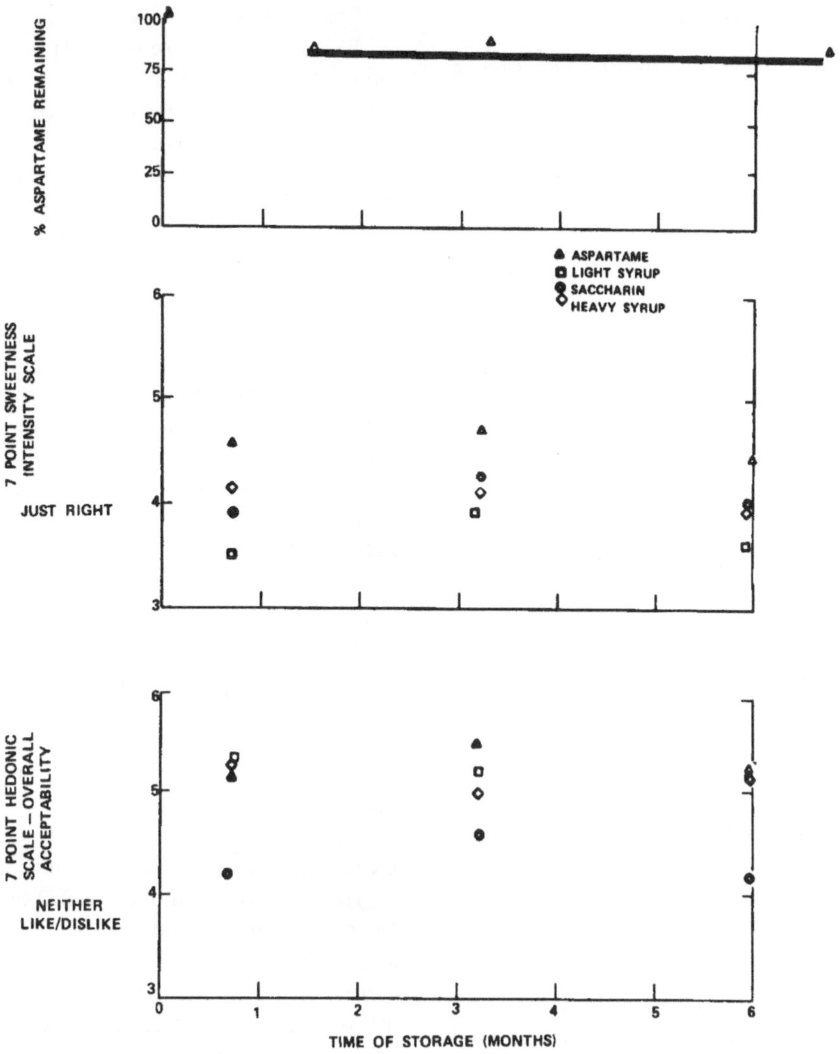


FIGURE 20. Storage stability study of aspartame-sweetened canned fruit cocktail (pH = 3.8). Top: chemical stability of aspartame; middle: stability of sweetness intensity; bottom: stability of acceptability. Storage was at room temperature.

One product that could easily be formulated is an aspartame-sweetened fruit juice which could be prepared simply by adding aspartame to the existing all natural concentrates. For orange juice concentrate at pH 3.8, a level of about 0.06% (0.015 in the prepared beverage) raises the sweetness to a more desirable level based on laboratory panels. A drink of this sort might require regulatory approval, as fruit juices are standardized and stringently regulated.

Preparation of a nonstandardized frozen fruit juice drink concentrate in which the final product contains 50% juice concentrates is also possible. The formula is shown in Table 35.

*4. Jams, Jellies, and Preserves*

A limited amount of work was done by a manufacturer of jam and jelly products; aspartame losses due to processing were not determined. Aspartame was used to sweeten dietetic jam and jelly at the level of 0.12%. Carrageenan and pectin were used to provide body; potassium sorbate was added as a preservative. Since sugar plays a role in the setting of certain grades of pectin and also has a preservative action by its osmotic effect, other ingredients must provide these functions.

Products were stored at 45 and 72°F. Organoleptic evaluation demonstrated that the high temperature caused perceived sweetness loss to occur between 4 and 8 months' storage, while the product stored at 45°F remained sweet after 11 months. Because of the high sugar content in jams, jellies, and preserves, substantial calorie reductions are possible.

The use of fruit preserves in commercial rather than retail channels may be more practical than their use for home preserving. The turnover of preserves suitable for use in yogurt and baking may be rapid enough to take advantage of aspartame-sweetened products. In addition, institutional portion controlled packs may be practical in controlled distribution systems.

TABLE 35

**Aspartame-sweetened Orange Breakfast Drink (Frozen Beverage Concentrate)**

| Ingredients | Concentrate (%) |
|---|---|
| Frozen orange juice concentrate | 49.0 |
| Solution containing: | |
| 1.5% Sodium carboxymethyl cellulose | 32.1 |
| 2.4% Citrus cloud | |
| Water | 9.62 |
| Orange pulp | 5.80 |
| Citric acid solution, 50% | 1.93 |
| Orange flavors | 0.65 |
| Yellow colors, 5% solutions | 0.34 |
| Trisodium phosphate | 0.27 |
| Potassium citrate | 0.14 |
| Ascorbic acid | 0.060 |
| Aspartame | 0.058 |
| Tricalcium phosphate | 0.041 |
| | 100 |

*Note:* Preparation – dilute with three equal volumes of water. Calories compared to unstandardized sugar counterpart = 60%. Calories compared to standardized unsweetened counterpart = 55%.

TABLE 36

**Aspartame Frozen Dessert**

| Ingredients | Weight (%) |
|---|---|
| Milk fat | 3.0 |
| Milk solids, nonfat | 11.65 |
| Sorbitol, dry | 7.75 |
| Stabilizer | 1.50 |
| Vanilla extract (4 X) | 0.10 |
| Aspartame | 0.06 |
| Color | 0.02 |

*Note:* 70% Overrun; 102.4 cal/100 g; standardized ice cream = 207 cal/100 g.

## G. Frozen Dessert and Dairy Products

Because of the neutral pH of many dairy products, aspartame may not seem compatible with these systems. However, yogurt is cultured to a pH that provides optimum stability for aspartame. It is kept in refrigerated storage and has an otherwise limited shelf life based on microbial stability. Aspartame can be added to the product just prior to packaging.

In frozen desserts, aspartame addition also follows pasteurization. These products are frozen and have an otherwise limited shelf life due to lactose crystallization, product shrinkage, and gumminess. These two systems provide some of the applications best suited to aspartame. Presently, very few products fill this class; saccharin is fairly incompatible with dairy products for organoleptic reasons.

### *1. Frozen Dessert*

As described earlier in the regulatory discussion, standards of identity preclude ice cream and ice milk development with any significant caloric reduction. However, an excellent aspartame-sweetened, frozen dairy dessert can be prepared comparing very well with standardized products and having only half the calories of ice cream. Table 36 shows the composition.

The product described was compared with three commercial products: ice cream, ice milk, and a dietetic frozen dessert. The tests were carried out as unidentified paired comparisons and 40 to 45 experienced but untrained judges participated. They were asked to rate sweetness, flavor, iciness, gumminess, overall texture, and overall acceptability, and were asked to indicate textural and overall preference. Ratings for the aspartame frozen dessert indicated it was more icy but less gummy than either the ice milk or diet dessert. Additionally, it was preferred over the ice milk for all quality attributes and over the dietetic product for overall acceptability. Though there were no significant differences in quality attributes between the aspartame product and the ice cream an overall preference was shown for ice cream.

Further product improvements are possible with the availability of a suitable bulking material.

### *2. Yogurt*

Yogurt demonstrated substantial market growth in the U.S. in recent years, possibly due to its good taste, low calorie content, and wholesome appeal. The calorie perceptions for fruit-filled products are simply untrue. However, with aspartame, yogurt can be low calorie. Swiss- or sundae style yogurt products comprise the bulk of yogurt sales and contain significant precentages of fruit preserves. These preserves increase the calories of plain, low fat yogurt (110 cal/8-oz cup) more than twice (260 to 300 cal/8-oz cup).

With aspartame, colored and flavored products can be made barely increasing the calories of plain, low fat yogurt. Also, replacing the sugared preserves with more yogurt raised the product's protein content and wholesomeness. Aspartame-sweetened yogurt is well received but may require fruit pieces for visual and textural flavor reinforcement. This can be done (with a slight price increase) by adding individually quick-frozen or unsweetened fresh fruit without significant caloric increases. On a volume basis, the caloric content of fruit and plain yogurt is about the same. A typical fruitless yogurt formula is shown in Table 37.

G. D. Searle & Co. conducted a storage stability study in which five yogurt flavors were tested for analytical aspartame recovery and for sensory evaluation of sweetness and overall acceptability. Total acidity, pH, and microbiological stability were also measured. Figure 21 shows that after 30 days of refrigerated storage, less than one fourth aspartame had been converted; sweetness perception changed little, and overall acceptability was stable. As expected, the pH dropped from 4.5 at the end of incubation to about 4.0 and gradually stabilized at pH 3.8. Correspondingly, the titratable acidity increased gradually. The products were microbiologically stable within generally acceptable limits of 8 weeks; the product was screened for yeasts and molds, coliforms, *E. coli,* coagulase positive *S. aureus,* and *Salmonella.*

The aspartame products were also compared to Swiss-style commercial products in unidentified, paired comparison tests. Sweetness and overall acceptability were judged by 30 to 35 experienced panelists. Of the three flavors tested, strawberry and pineapple were not significantly different from the commercial counterparts containing sugar. The cherry-flavored yogurt sweetened with aspartame was preferred significantly for overall acceptability, but because it was also significantly sweeter, the result is uninterpretable. It should be noted that these results occurred when the aspartame product contained no fruit pieces. The conventional products contained fruit.

TABLE 37

**Aspartame-sweetened Blueberry Yogurt**

| Ingredients | Amount |
|---|---|
| Plain yogurt base[a] | 100 gal |
| Aspartame[b] | 0.56 lb |
| Flavor | 4.40 lb |
| Color | 0.39 lb |

[a]Cup incubation is inappropriate when aspartame is used in yogurt manufacture.
[b]To avoid sweetness loss, the aspartame, flavor, and color should be added after the incubated batch is cooled to 80–85°F.

### H. Other Applications

The preceding discussion covered in detail a wide range of products. A multitude of product opportunities exist for aspartame use in low calorie and dietetic foods, but this chapter is limited to discussion of completed work in established food categories.

Peanut butter is difficult to sweeten conventionally without violating the standard identity relative to a high peanut content. Individually quick-frozen fruit could be sweetened without the osmotic effects of sugar. Surprisingly, aspartame is claimed to prevent oxidation in fats, oils, and emulsions at 0.01 to 0.1% of the lipid.[154] Aspartame has also been used to prepare an anhydrous dentifrice composition.[155] Low calorie honey or pancake syrup can be formulated. The data presented here provide basic concepts from which many new products may be created.

## IX. CLINICAL SIGNIFICANCE

Marketing reserach data compiled in 1975 by G. D. Searle & Co. indicated that 76% of women feel they are overweight, and 50% of the respondents in this study said that they had dieted during the past year. Of those who dieted, 80% failed to achieve their weight-loss goals.

It has been established that even without aspartame, there is already a very substantial market for calorie-reduced foods and that these foods are often sweetened with saccharin. Should aspartame be cleared for use alone or in combination with other sweeteners, it should be possible to improve the acceptability of a large number of

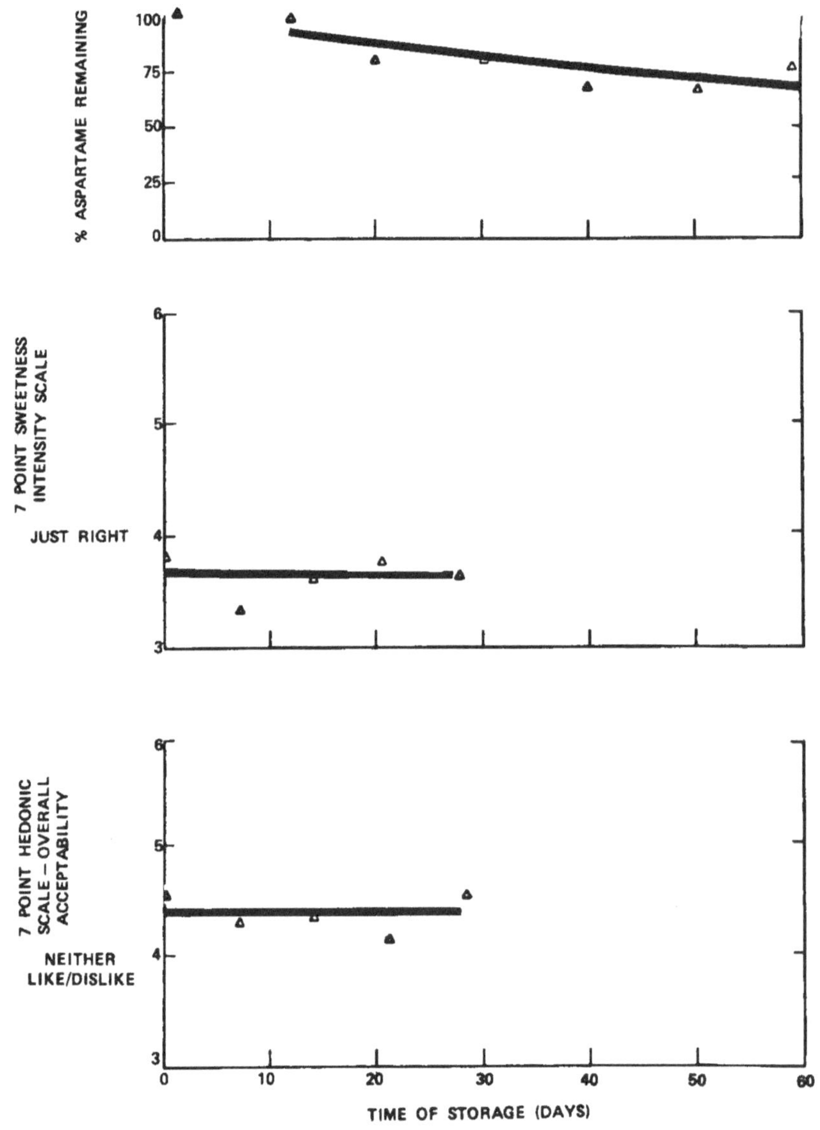


FIGURE 21. Storage stability of an aspartame-sweetened, blueberry-flavored yogurt (pH = 3.8). Top: chemical stability of aspartame; middle: stability of sweetness intensity; bottom: stability of acceptability. Samples were stored refrigerated.

calorie-reduced foods and perhaps further expand the market. It is questionable whether or not calorie-reduced foods result in weight control or loss. Do the obese eat two servings of a calorie-reduced product?

A great deal of literature discussing the psychological profile of obese persons is available. Chase[156] and Schachter[157,158] have published significant studies. Schachter's studies indicate that the moderately obese respond strongly to external stimuli. One could interpret this to mean that highly acceptable, low calorie products could be substituted in the diet without the individual compensating with other calorie sources.

Porikos et al.[159] have recently carried out preliminary studies in which the diets of eight patients were controlled by including foods that could be presented as standard menu items containing sugar or as calorically reduced covert substitutions sweetened with aspartame. The patients were provided with the sugar-based menu for 3 days (as a baseline); then they were fed a calorically reduced diet for 6 days. During the final 6 days, sugar-sweetened foods were again included in meals. The calorie-reduced products containing aspartame were equally accepted by the patients and were not perceived as calorie-reduced. In addition, the *ad libitum* caloric intake of patients on the calorie-reduced diets was 25% below intake of patients on the control diet. There were no clear-cut indications of calorie compensation by increases in food consumption; however, this study served mainly to establish the utility of a good tasting sweetener. Lacking a sweetener with aspartame's taste quality, studies like this would not be possible.

An earlier study by Campbell et al.[160] indicated that via a "feeding machine" lean, young adults regulate caloric intake regardless of caloric density; grossly obese adults do not. In that study, a drink's caloric density was varied by dilution, and the volumes taken were measured over several months. Other workers, including Jordan,[161] Wooley,[162] and Spiegel[163] have done work relating diet regulation to caloric intake and obesity. Results overall are not clear-cut.

There is critical need for a major study of weight control through diet, and studies cited above show that data are obtainable. A study should employ a large sample size from various weight groups, ages, etc., and expose them to various diets and controls over extended time periods in realistic settings. The expense of such a study might require cooperative industrial or government funding. The study's importance to national health would justify it.

## CONCLUSIONS

Using sophisticated technology, a multitude of highly acceptable, calorie-reduced foods can be prepared. The scope and quality of these diet foods depend on regulatory factors and on the quality and availability of intense sweeteners and bulking agents. High potency sweeteners are only one of several important factors in an overall program of preparing calorie-reduced foods. Aspartame is a high potency sweetener suitable for application in a wide range of foods. Its good quality sweet taste makes it unique when compared to other intense sweeteners. Aspartame can be blended with other nutritive and nonnutritive sweeteners for flavor improvement and economic benefits. That highly acceptable, low calorie foods are effective in weight control is unproven and deserves further investigation. Methodology to do so is available.

**Acknowledgments**

This chapter was prepared with the assistance of a great number of people both inside and outside of G. D. Searle & Co. The author is particularly grateful to John Colliopoulos and Louis Jokay who developed most of the aspartame applications; Teresita Muzones who searched out much of the literature; and Geraldine Peters and Annette Ripper who assisted in the laborious details of assembling, proofreading, and editing the manuscript. William Colaianni played a major role in constructing Table 4. Kathleen Wolfe participated in designing and analyzing many of the sensory studies. Some of the data presented were performed on contract, and other data were generously supplied by researchers in the food industry.

## REFERENCES


1. **Lundgren, A. O. and Ramsey, G.,** *Dietetic Foods 1974,* Charles H. Kline, Fairfield, N.J., 1974, 7.
2. **Abrams, I. J.,** *Fifth National Household Menu Census Study,* Market Research Corporation of America, Chicago, Ill., 1975.
3. **Page, L. and Friend, B.,** in *Sugars in Nutrition,* Sipple, H. L. and McNutt, K. W., Eds., Academic Press, New York, 1974, 102.
4. **Yudkin, J.,** *Sweet and Dangerous,* Peter H. Wyden, New York, 1972.
5. **Dufty, W.,** *Sugar Blues,* Chilton Book, Radnor, Penn., 1975, 51.
6. **Brown, K.,** G. D. Searle & Co., personal communication, August 10, 1973.
7. **Stare, F. J.,** Sugar in the diet of man, *World Rev. Nutr. Diet.,* 22, 1975, 239.
8. **Sipple, H. L. and McNutt, K. W.,** *Sugars in Nutrition,* Academic Press, New York, 1974.
9. **Painter, N. S. and Burkitt, D. P.,** Diverticular disease of the colon, a 20th century problem, *Clin. Gastroenterol.,* 4, 3, 1975.
10. **Burkitt, D. P., Walker, A. R. P., and Painter, N. S.,** Dietary fiber and disease, *JAMA,* 229, 1068, 1974.
11. **Spillar, G. A. and Amen, R. J.,** Dietary fiber in human nutrition, *CRC Crit. Rev. Food Sci. Nutr.,* 7, 39, 1976.
12. **Tatem, J. W., Jr.,** Speech presented at Symposium on Sugars in Nutrition, Chicago Nutrition Association, January 19, 1976.
13. Informatics, Inc., Scientific Literature Reviews on (GRAS) Food Ingredients-Sucrose, National Technical Information Service, U.S. Department of Commerce, December 1973.
14. **Campbell, G. D.,** testimony, The Role of Sucrose in the Precipitation and Emergence of Diabetes, before the U.S. Senate Select Committee On Nutrition and and Human Needs, May 1, 1973.
15. **Dickens, F., Randle, P. J., and Whelan, W. J.,** *Carbohydrate Metabolism and Its Disorders,* Academic Press, New York, 1970.
16. *Diabetes Literature Index,* National Institute of Arthritis, Metabolism and Digestive Diseases and American Diabetes Association, U.S. Government Printing Office, Washington, D.C., 1974, 7.
17. *Exchange Lists for Meal Planning,* Committee on Foods and Nutrition, American Diabetes Association, Inc., 1976.
18. **Bowen, W. H.,** Physiological effects of food carbohydrates, in *ACS Symposium Series,* Jeanes, A. and Hodge, J., Eds., American Chemical Society, Washington, D.C., 15, 1975, 150.
19. **Newbrun, E.,** Diet and dental caries relationships, in *Advances in Caries Research,* 2, 1974, xv.
20. **Finn, S. B. and Glass, R. B.,** Sugar and dental decay, *World Rev. Nutr. Diet.,* 22, 304, 1975.
21. **Anon.,** Giant Food considers sugar labeling for pre-sweetened cereals, *Food Chem. News,* 18(10), 10, 1976.
22. **Anon.,** Gallup poll finds increased consumer concern with food additives, *Food Chem. News,* 18(10), 45, 1976.
23. National Food Situation, U.S. Department of Agriculture, Economic Research Service, U.S. Government Printing Offices, Washington, D.C., 154, 28, November, 1975.
24. Sugar and Sweetener Report, U.S. Department of Agriculture, Agricultural Marketing Service, U.S. Government Printing Offices, Washington, D.C., 1(3), 11, April, 1976.
25. **Becker, R. H.,** Sweeteners – a study in regulatory lore, *Food Prod. Dev.,* 9(9), 26, 1975.
26. **Ebert, A. G.,** Advice to the technologist developing food additives, *Food Prod. Dev.,* 10(5), 40, 1976.
27. **Anon.,** FDA asks firms to submit letter on saccharin-sorbitol frozen dessert questions, *Food Chem. News,* 18(9), 24, 1976.
28. **Berliner, J.,** G. D. Searle & Co., personal communication, 1972.
29. **Rennhard, H. H.,** Dietetic foods, U.S. Patent 3,876,794, Pfizer, Inc., Groton, Conn., April 8, 1975.
30. **Naito, F.,** The properties and applications of malbit, *New Food Industry,* 13(9), 1, 1971.
31. **Scieweck, H., Steinle, G., and Haberl, L.,** Low calorie sweetener and sweetener base, U.S. Patent 3,865,957, Sueddeutsche Zucker-Aktiengesellschaft, Mannheim, W. Germany, February 11, 1975.
32. **Mattson, F. H. and Volpenhein, R. A.,** Low calorie fat-containing food composition, U.S. Patent 3,600,186, Proctor & Gamble Co., Cincinnati, Oh., August 17, 1971.
33. **Fallat, R. W., Glueck, C. J., Mattson, F., Ishikawa, T. T., and Brazier, J.,** Sucrose polyester: effect on fecal cholesterol, total neutral sterol, and fatty acids, *Clin. Res.,* 24, 360A, 1976.
34. **Mattson, F. H. and Nolan, G. A.,** Absorbability by rats of compounds containing from one to eight ester groups, *J. Nutr.,* 102, 1171, 1972.
35. **Fallat, R. W., Glueck, C. J., Lutmer, R., and Mattson, F. H.,** Short term study of sucrose polyester, a non-absorbable fat-like material, as a dietary agent for lowering plasma cholesterol, accepted for publication by *Am. J. Clin. Nutr.,*
36. **Mattson, F. H., Jandacek, R. J., and Webb, M. R.,** The effect of a nonabsorbable lipid, sucrose polyester, on the absorption of dietary cholesterol by the rat, *J. Nutr.,* 106(6), 747, 1976.
37. **Rodgers, J. B. and O'Conner, P. J.,** Effect of phosphatidylcholine on fatty acid and cholesterol absorption from mixed micellar solutions, *Biochim. Biophys. Acta,* 409, 192, 1975.
38. **Gerskowitch, V. P. and Russel, R. I.,** Tritiated glycerol triether as an oil phase marker in man, *J. Lipid Res.,* 15, 432, 1974.
39. **Morgan, R. G. and Hofmann, A. F.,** Oil-phase marker, to estimate fat absorption in rats with cholestyramine-induced steatorrhea, *J. Lipid Res.,* 11, 231, 1970.

40. Anon., GAO calls on FDA to reassess interim order for saccharin, *Food Chem. News,* 18(26), 60, 1976.
41. **Daniels, R.,** *Sugar Substitutes and Enhancers,* Noyes Data Corp., Park Ridge, N.J., 1973, 80.
42. **Larson, N. L.,** Sensory Properties of Flavored Gelatins Containing Sucrose or Synthetic Sweeteners, M.Sci. thesis, University of California, Davis, Cal., 1975.
43. **Salant, A.,** Nonnutritive sweeteners, in *Handbook of Food Additives,* Furia, T. E., Ed., The Chemical Rubber Co., Cleveland, 1972, chap. 13.
44. **Mazur, R. H.,** Aspartic acid-based sweeteners, in *Symposium: Sweeteners,* Inglett, G. E., Ed., AVI, Westport, Conn., 1974, chap. 14.
44a. *Federal Register,* 40(235), 1975, p. 56907.
45. **Inglett, G. E.,** *Recent Sweetener Research,* Botanicals, Peoria, Ill., 1971.
46. **Williams, G.,** Sugar is sweet, but financial rewards for finding a low calorie substitute are even sweeter . . . , *Sci. Dig.,* 41, August, 1975.
47. Symposium on Sugar and Sugar Replacements – Symposium Proceeding No. 23, British Food Manufacturing Industries Research Association, London, 1974.
48. **Russo, J. R.,** Xylitol: anti-carie sweetener?, *Food Eng.,* 48(4), 77, 1976.
49. **Hodge, J. E. and Inglett, G. E.,** Structural aspects of glycosidic sweeteners containing (1′-2)-linked disaccharides, in *Symposium: Sweetners,* Inglett, G. E., Ed., AVI, Westport, Conn., 1974, chap. 20.
50. **Birch, G. G. and Lee, C. K.,** The chemical basis of sweetness in model sugars, in *Sweetness and Sweeteners,* Birch, G. G., Green, L. F., and Coulson, C. B., Eds., Applied Science, London, 1971, 95.
51. **Brussel, L. P., Peer, H. G., and Vanderhejden, A.,** Structure-taste relationship of some sweet tasting dipeptide esters, *Z. Lebensm. Unters. Forsch.,* 159, 337, 1975.
52. **Huber, U. A.,** Sweetness in human nutrition (German), *Neue Zuricher Zeitung,* 131, 24, 1974.
53. **Newbrun, E.,** Sugar, sugar substitutes and noncaloric sweetening agents, *Int. Dent. J.,* 23, 328, 1973.
54. **Beck, K. M.,** Synthetic sweeteners, past present and future, in *Symposium: Sweeteners,* Inglett, G. E., Ed., AVI, Westport, Conn., 1974, chap. 11.
55. **Dermer, O.,** The science of taste, *Proc. Okla. Acad. Sci.,* 27, 9, 1946.
56. **Cook, M. K. and Gominger, B. H.,** Glycyrrhizin, in *Symposium: Sweeteners,* Inglett, G. E., Ed., AVI, Westport, Conn., 1974, 211.
57. **Clauss, K. and Jensen, H.,** Oxathiazinone dioxides – a new group of sweetening agents, *Angew. Chem.,* 12, 869, 1973.
58. **Mazur, R. H.,** Aspartame – a sweet surprise, *J. Toxicol. Environ. Health,* 2(243), 1976.
59. **Beck, C. I.,** Sweetness, character and applications of aspartic acid based sweeteners, in *Symposium: Sweeteners,* Inglett, G. E., Ed., AVI, Westport, Conn., 1974, 164.
60. **Farnsworth, N. R.,** Current status of sugar substitutes, *Cosmet. Perfum.,* 88, 27, 1973.
61. **Horowitz, R. M. and Gentili, B.,** Dihydrochalcone derivatives and their use as sweetening agents, U.S. Patent 3,087,821, April 30, 1963.
62. **Horowitz, R. M. and Gentili, B.,** Flavonoid constituents of citrus, in *Citrus Processing Science and Technology,* Vol. I, Nagy, S., Shaw, P., and Veldhois, M. K., Eds., AVI, 1977, in press.
63. **Garbrecht, W. L.,** Substituted tetrazole, U.S. Patent 3,515,727, Eli Lilly & Co., Indianapolis, Ind., June 2, 1970.
64. **Garbrecht, W. L.,** Novel substituted tetrazole and use thereof, U.S. Patent 3,597,234, Eli Lilly & Co., Indianapolis, Ind., August 3, 1971.
65. **Cagan, R. H.,** Chemostimulatory protein: a new type of taste stimulus, *Science,* 181, 32, 1973.
66. **Selzer, R. J.,** Work on new synthetic sweeteners advances, *Chem. Eng. News,* p. 27, August 25, 1975.
67. **Morris, J. A. and Cagan, R. H.,** Purification of monellin, the sweet principle of *Dioscoreophyllum cumminsii, Biochim. Biophys. Acta,* 261, 114, 1972.
68. **Acton, E. M., Lerom, M. W., and Stone, H.,** 1,4-cyclohexadiene-1-carboxaldehyde *syn*-oxime synthetic sweetening agents, U.S. Patent 3,952,114, Stanford Research Institute, Menlo Park, Calif., April 20, 1976.
69. **Acton, E. M. and Stone, H.,** Potential new artificial sweetener from study of structure-taste relationships, *Science,* 193, 586, 1976.
70. **Neely, J. S. and Thompson, J. A.,** Composition of matter containing paramethoxycinnamaldehyde as a flavoring agent and sweetener, U.S. Patent 3,867,557, Proctor & Gamble Co., Cincinnati, Oh., February 18, 1975.
71. **Neely, J. S. and Thompson, J. A.,** Soft drink composition containing paramethoxycinnamaldehyde as a flavoring agent and sweetener, U.S. Patent 3,908,028, Proctor & Gamble Co., Cincinnati, Oh., September 23, 1975.
72. **Acton, E. M., Stone, H., Leaffer, M. A., and Oliver, S. M.,** Perillartine and some derivatives: clarification of structures, *Experientia,* 26, 473, 1970.
73. **Cagan, R. H.,** Biochemistry of sweet sensation, in *Sugars in Nutrition,* Academic Press, 1974, chap. 3.
74. **Crosby, G. A. and Saffron, P. M.,** Sweetener, U.S. Patent 3,952,058, Dynapol, Palo Alto, Cal., April 20, 1976.
75. **Nofre, C., Sabadie, J., and Bal, D.,** Sweetening properties of L-alanine *tert*-butyl ester, *Naturwissenschaften,* 61, 404, 1974.
76. **Chibata, I., Miyoshi, M., Ito, H., Fujei, T., and Kawashima, K.,** Aspartyl amide sweetening agents, U.S. Patent 3,971,822, Tanabe Seiyaku Co., Ltd., Osaka, Japan, July 27, 1976.
77. Takeda Chemical Industries, Ltd., Tokyo, Japan, L-aspartylaminomalonic acid diesters, British Patent 1,434,043, April 28, 1976.

78. **Fujino, M., Wakimasu, M., and Tanaka, K.**, L-aspartylaminomalonic acid diesters: a new group of compounds with intense sweetness, *Naturwissenschaften*, 60, 351, 1973.
79. **Nakjima, N., Aoko, H., Fujino, M., Nishimura, O., Wakimasu, M., and Mano, M.**, L-aspartyl aminomalonic acid methyl 2-methyl cyclohexyldiester, U.S. Patent 3,959,245, Takeda Chemical Industries, Ltd., Tokyo, Japan, May 25, 1976.
80. **Gregory, H.**, Sweetened compositions, British Patent 1,233,216, Imperial Chemical Industries Ltd., London, England, May 26, 1971.
81. **Schlatter, J.**, Optionally substituted aspartyl cyclohexylamine lower alkyl esters, compositions and method, U.S. Patent 3,714,139, G. D. Searle & Co., Skokie, Ill., January 30, 1973.
82. **Fujimoto, Y.**, Application of a new sweetener, Japanese Patent 7,308,829, Kyowa Hakko Kogyo Co., Ltd., Tokyo, Japan, March 17, 1973.
83. **Anon.**, Sucrose studies yield new supersweetener, *Chem. Eng. News*, p. 30, September 20, 1976.
84. **Tahara, A., Nakata, T., and Ohtsuka, Y.**, New type of compound with strong sweetness, *Nature*, 233, 619, 1971.
85. **Neeley, J. S.**, Compositions of matter containing heliotropyl nitrile as a flavoring agent and sweetener, U.S. Patent 3,778,517, Proctor & Gamble Co., Cincinnati, Oh., December 11, 1973.
86. **Windgassen, R. J.**, Artificial sweetening agent, U.S. Patent 3,340,070, Shell Oil Co., Houston, Tex., September 25, 1967.
87. **Crosby, G. A., Dubois, G. E., and Weinshenker, N. M.**, Ionic sweetener, U.S. Patent 3,974,299, Dynapol, Palo Alto, Cal., August 10, 1976.
88. **Lapidus, M. and Sweeney, M.**, L-4′cyano-3-(2,2,2-Trifluoracetamido) succinanilic acid and related sweetening agents, *J. Med. Chem.*, 16, 163, 1973.
89. **Zaffaroni, A.**, Non-absorbable non-nutritive sweeteners, U.S. Patent 3,876,816, Dynapol, Palo Alto, Cal., April 8, 1975.
90. **Van der Wel, H. and Loeve, K.**, Isolation and characterization of thaumatin I and II, the sweet-tasting protein from *Thaumatococcus daniellii*, Benth., *Eur. J. Biochem.*, 31, 221, 1972.
91. **Bartoshuk, L. M., Lee, C., and Scarpellino, R.**, Sweet taste of water induced by artichoke (*Cynara scolymus*), *Science*, 178, 988, 1972.
92. **Finley, J. W.**, Kynurenine derivatives as sweetening agents, U.S. Patent 3,702,255, the United States of America as represented by the Secretary of Agriculture, November 7, 1972.
93. **Finley, J. W. and Friedman, M.**, New sweetening agents: *N′*-formyl-and *N′*-acetyl-kynurenine, *J. Agric. Food Chem.*, 21, 33, 1973.
94. **Lee, Chi-Hang**, Intense sweetener from Lo Han Kuo (*Momordica grosvenori*), *Experientia*, 31, 533, 1975.
95. **Jizba, J., Herout, V., Sorm, F., and Dolejs, L.**, The structure of osladin – the sweet principle of *Polypodium vulgare* L, *Tetrahedron Lett.*, 18, 1329, 1971.
96. **Dorfman, R. I. and Nes, W. R.**, Antiandrogenic activity of dihydroisosteviol, *Endocrinology*, 67, 282, 1960.
97. **Kibun, K. K.**, Natural sweetening agent stevioside manufacture by plant tissue culture, Japanese Patent 1,019,169, August 5, 1974.
98. **Parli, C. J., Kriefer, P., and Schmidt, B.**, 4-Chloro-3-hydroxyanthranilic acid (ClHaa) – a potent irreversible inhibitor of liver and kidney 3-hydroxy-anthranilic acid oxidase (HAAO), *Pharmacologist*, 17, 250, 1975.
99. **Kornfeld, E. C., Sheneman, J. M., and Suarez, T.**, Nonnutritive sweeteners, German Offenfundig 1,917,844, Eli Lilly & Co., Indianapolis, Ind., November 6, 1969.
100. **Scott, D.**, Aspartame: properties, use and present status, presented at IVth International Congress of Food Science and Technology, Spain, 1974.
101. **McCormick, R. D.**, Aspartame: a new dimension for controlling product sweetness, *Food Prod. Dev.*, 9(1), 22, 1975.
102. Equa 200: Brand of Aspartame, A New Food Ingredient Sweetener, G. D. Searle & Co., Skokie, Ill., 1974.
103. **Mazur, R. H., Goldkamp, A. H., and Schlatter, J. M.**, Aspartic acid containing dipeptide lower alkyl esters, U.S. Patent 3,475,403, G. D. Searle & Co., Skokie, Ill., October 28, 1969.
104. **Mazur, R. H., Schlatter, J. M., and Goldkamp, A. H.**, Synthesis and taste of dipeptide esters, *J. Am. Chem. Soc.*, 91, 2684, 1969.
105. **Ariyoshi, Y. and Sato, N.**, The separation of $\alpha$ and $\beta$-L-aspartyl-L-phenylalanine methyl ester with aromatic acids, *Bull. Chem. Soc. Jpn.*, 45, 942, 1972.
106. Code of Federal Regulations, U.S. Government Printing Office, Washington, D.C., 21 Sec. 121.1258, April, 1975.
107. **Furda, I., Malizia, P. D., Kolor, M. G., and Vernieri, P. J.**,Decomposition products of L-aspartyl-L-phenylalanine methyl ester and their identification by gas-liquid chromatography, *J. Agric. Food Chem.*, 23, 340, 1975.
108. **Ariyoshi, Y., Yamatani, T., Uchiyama, N., Yasuda, N., Toi, K., and Sato, N.**, Method of producing $\alpha$-L-aspartyl-L-phenylalanine alkyl esters, U.S. Patent 3,833,553, Ajinomoto Co., Inc., Tokyo, Japan, September 3, 1974.
109. **Ariyoshi, Y. and Toi, K.**, Method of purifying $\alpha$-L-aspartyl-L-phenylalanine lower alkyl esters of peptide impurities, U.S. Patent 3,833,554, Ajinomoto Co., Inc., Tokyo, Japan, September 3, 1974.
110. **Bachman, G. L., Ottedahl, M. L., and Vineyard, B. D.**, Process for the preparation of $\alpha$-L-aspartyl-L-phenylalanine alkyl esters, U.S. Patent 3,933,781, Monsanto Co., St. Louis, Mo., January 20, 1976.

111. **Cloninger, M. R. and Baldwin, R. E.,** Aspartylphenylalanine methyl ester: a low calorie sweetener, *Science,* 170, 81, 1970.
112. **Cloninger, M. R. and Baldwin, R. E.,** L-Aspartyl-L-phenylalanine methyl ester (aspartame) as a sweetener, *J. Food Sci.,* 39, 347, 1974.
113. **Moskowitz, H. R. and Dubose, C.,** Taste intensity, pleasantness and off-taste of aspartame, sugars and their mixtures, *J. Food Sci.,* submitted for publication.
114. **Eisenstadt, M. E.,** Calorie-free dipeptide sweetening composition, U.S. Patent 3,875,311, Cumberland Packing Corporation, Bracebridge, Ont., Canada, April 1, 1975.
115. **Eisenstadt, M. E.,** Dipeptide sweetener composition, U.S. Patent 3,875,312, Cumberland Packing Corp., April 1, 1975.
116. **Schade, H. R.,** Taste modifier for artificial sweetners, U.S. Patent 3,934,047, General Foods Corp., White Plains, N.Y., January 20, 1976.
117. **Bahoshy, B. J., Klose, R. E., and Nordstrom, H. A.,** Chewing gums of longer lasting sweetness and flavor, U.S. Patent 3,943,258, General Foods Corp., White Plains, N.Y., March 9, 1976.
118. **Finucane, T. P. and Halik J. J.,** Dipeptide low grade coffees, U.S. Patent 3,829,588, General Foods Corp., White Plains, N.Y., August 13, 1974.
119. General Foods Corp., White Plains, N.Y., Process for enhancing fruit flavors, U.S. Patent Application 1418-087, December 17, 1975.
120. **Pangborn, R. M., Trabue, I. M., and Szczesniak, A. S.,** Effect of hydrocolloids on oral viscosity and basic taste intensities, *J. Texture Stud.,* 4, 220, 1973.
121. **Scott, D.,** Sweetening composition and method for use thereof, U.S. Patent 3,780,189, G. D. Searle & Co., Skokie, Ill., December 18, 1973.
122. **Yamaguchi, S., Shimizu, A., Kirimura, J., and Ninomiya, T.,** Method of improving the taste of artificial sweeteners, Japanese Patent 7,223,389, Ajinomoto Co., Inc., Tokyo, Japan, June 26, 1972.
123. **Wolfe, K. A. and Mattes, K.,** Synergistic effects of 1-methyl-*N*-L-aspartyl-L-phenylalanine (aspartame) with sucrose and glucose, *J. Food Sci.,* submitted for publication.
124. **Moskowitz, H. R.,** Models of additivity for sugar sweetness, in *Sensation and Measurement,* D. Reidel, Dordrecht, Germany, 1974, 379.
125. **Moskowitz, H. F.,** Models of sweetness additivity, *J. Exp. Psychol.,* 99, 88, 1973.
126. **Wardrip, E. K.,** Sweeter corn syrups may be used in place of surcrose and invert syrups, *Food Technol.,* 25, 47, 1971.
127. **Stone, H., Oliver, S., and Kloehn, J.,** Temperature and pH effects on the relative sweetness of suprathreshold mixtures of dextrose, fructose, *Percept. Psychophys.,* 5(5), 257, 1969.
128. **LaVia, A. L. and Hill, J. A.,** Sweetening compositions, U.S. Patent Application B439,788, E. R. Squibb & Sons, Inc., Port Chester, N.Y., February 3, 1976.
129. **Stone, H. and Oliver, S. M.,** Measurement of relative sweetness of selected sweeteners and sweetener mixtures, *J. Food Sci.,* 34, 215, 1969.
130. **Ziemba, J. V.,** What to know about non-nutritive sweetener blends, *Food Eng.,* 41(5), 82, 1969.
131. **Shoaf, M. D. and Pischke, L. D.,** Sweetening composition and process therefore, U.S. Patent 3,928,633, December 23, 1975.
132. American College of Nutrition Proceedings, Scientific review of a new sweetener, First annual interim meeting, Chicago, 1974, *J. Toxicol. Environ. Health,* 2(1) and (2), 1976.
133. **Oppermann, J. A., Muldoon, E., and Ranney, R. E.,** Metabolism of aspartame in monkeys, *J. Nutr.,* 103, 1454, 1973.
134. **Opperman, J. A., Muldoon, E., and Ranney, R. E.,** Effect of aspartame on phenylalanine in the monkey, *J. Nutr.,* 103, 1460, 1973.
135. **McCormick, R. D.,** Extensive test program confirms safety for aspartame, new dipeptide sweetener, *Food Prod. Dev.,* December, 8(10), 32, 1974.
136. G. D. Searle & Co., Skokie, Ill., Master file #134 for Food Additive Petition 3A,2885, filed with the U.S. Food and Drug Administration, February 27, 1973.
137. **Furda, I. and Trumbetas, J. F.,** Multifunctional sweetening composition, U.S. Patent 3,934,048, General Foods Corp., White Plains, N.Y., January 20, 1976.
138. **Cella, J. A. and Schmitt,W. H.,** Low calorie dry sweetener composition, U.S. Patent 3,753,739, Alberto Culver Co., Melrose Park, Ill., August 21, 1973.
139. **Glicksman, M. and Wankier, B. N.,** Low calorie sweetening composition and method for making same, U.S. Patent 3,922,369, General Foods Corp., White Plains, N.Y., November 25, 1975.
140. **Glicksman, M. and Wankier, B. N.,** Method for making a low calorie sweetening composition, U.S. Patent 3,761,288, General Foods Corp., White Plains, N.Y., September 25, 1973.
141. **Berg, G. H. and Trumbetas, J. F.,** Method of producing fast-dissolving dipeptide sweetener for sweet-sour systems, U.S. Patent 3,868,472, General Foods Corp., White Plains, N.Y., February 25, 1975.
142. **Shoaf, M. D. and Pischke, L. D.,** Stabilized methyl L-aspartyl-L-phenylalanine preparation by heating with a matrix forming material and surfactant, then grinding, U.S. Patent 3,956,507, General Foods Corp., White Plains, N.Y., May 11, 1976.

143. **Mazur, R. H.**, Aspartame – a sweet surprise, *J. Toxicol. Environ. Health*, 2(1), 243, 1976.
144. **Pischke, L. D. and Shoaf, M. D.**, Dry free-flowing stable and soluble synthetic sweeteners containing particulate L-aspartic acid dipeptide derivatives in a covered matrix, U.S. Patent 3,962,468, General Foods Corp., White Plains, N.Y., June 8, 1976.
145. General Foods Corp., White Plains, N.Y., Dipeptide salts – sweeteners, British Patent application 1431–057, April 7, 1976.
146. **Shoaf, M. D. and Pischke, L. D.**, Spray-dried aspartic acid derivatives, U.S. Patent Application 488,845, March 7, 1974.
147. General Foods Corp., White Plains, N.Y., Cold-water soluble aspartic acid sweeteners prepared by forming a paste with an inorganic salt then drying, U.S. Patent Application 477,569, October 6, 1974.
148. **Hornyak, J. and Stahl, H. D.**, Dry carbonation source for a beverage concentrate and a method of preparing same, U.S. Patent 3,939,289, General Foods Corp., White Plains, N.Y., February 17, 1976.
149. **Moskowitz, H. R.**, Applications of psychophysics and psychometrics to food evaluation. II. Ratio scaling, *Lebensm. Wiss. Technol.*, 8, 249, 1975.
150. Tanabe Seiyaku Co., Ltd., Tokyo, Japan, Food flavor improvement – using addition of L-aspartyl-L-phenylalanine methyl ester, Japanese Patent 29146, September 7, 1973.
151. **Baggerly, P. A.**, Cereal process and production, U.S. Patent 3,955,000, General Foods Corp., White Plains, N.Y., April 5, 1975.
152. **Rousseau, P. M.**, Dipeptide sweetening of comestibles, U.S. Patent 3,947,600, General Foods Corp., White Plains, N.Y., March 30, 1976.
153. Confectionery Manufacturers' Sales and Distribution, 1973, U.S. Department of Commerce, National Technical Information Service, Springfield, Va., June, 1974.
154. Ajinomoto Co. Inc., Tokyo, Japan, Preventing oxidtion of oils and fats, Japanese Patent 7,127,473, August 10, 1971.
155. **Barth, J. B.**, Anhydrous dentifrice composition – containing a dipeptide sweetener in combination with an acid anhydride, U.S. Patent 3,939,261, Colgate Palmolive Co., Piscataway, N. J., February 17, 1976.
156. **Chase, J.**, Obese in the land of milk and honey, *Hum. Behv.*, 4(6), 56, 1975.
157. **Schachter, S.**, Obesity and eating, *Science*, 161, 751, 1968.
158. **Schachter, S.**, Some extraordinary facts about obese humans and rats, *Am. Psychol.*, 26(2), 129, 1971.
159. **Porikos, K. P., Booth, G., and Van Itallie, T. B.**, Effect of Covert Nutritive Dilution on the Spontaneous Food Intake of Obese Individuals: A Pilot Study, *Am. J. Clin. Nutr.*, 30, 1638, 1977.
160. **Campbell, R. G., Hashim, S. A., and van Itallie, T. B.**, Studies of food intake – regulation in man: responses to variations in nutritive density in lean and obese subjects, *N. Engl. J. Med.*, 285, 1402, 1971.
161. **Jordan, H. A.**, Voluntary intragastric feeding: oral and gastric contributions to food intake and hunger in man, *J. Comp. Physiol. Psychol.*, 68, 498, 1969.
162. **Wooley, O. W.**, Long term food regulation in the obese and nonobese, *Psychosom. Med.*, 33, 436, 1971.
163. **Spiegel, T. A.**, Caloric regulation of food intake in man, *J. Comp. Physiol. Psychol.*, 84, 24, 1973.

Chapter 6

# PURE FRUCTOSE AND ITS APPLICATIONS IN REDUCED-CALORIE FOODS

**T. F. Osberger and H. R. Linn**

## TABLE OF CONTENTS



## I. INTRODUCTION

During the last decade, European developments in sucrochemical technology have made pure crystalline fructose available for industrial use.[2] In the creation of specialty foods, this new technology facilitates use of the unique monosaccharide, fructose, to achieve specific sweetening objectives. When formulating foods with special nutritional or dietary requirements, the availability of suitable carbohydrates to fulfill these requirements is highly desirable.

The food technologist engaged in formulating products labeled "dietetic" or "reduced calories" is soon obliged to face the reality of human propensity for consuming sweets.[5,25] Otherwise, no matter how appealing the product label initially appears to calorie conscious individuals, most likely the product will eventually fail on the market as it will not satisfy minimum consumer sweetness demands. Too often, consumers' previous dietary preferences for regular-calorie foods are ignored, and new products are formulated with the austerity of the eventual nutritional label as the prime objective. This misses the point – the typical dieting person wishes to reduce caloric intake by an effective regime demanding least restriction of former eating pleasures. Consequently, when selecting a sweetening system for dietetic or reduced-calorie products, it is important to choose those ingredients providing maximum sweetness for a given caloric expenditure.

TABLE 1

**Anomers of Fructose**

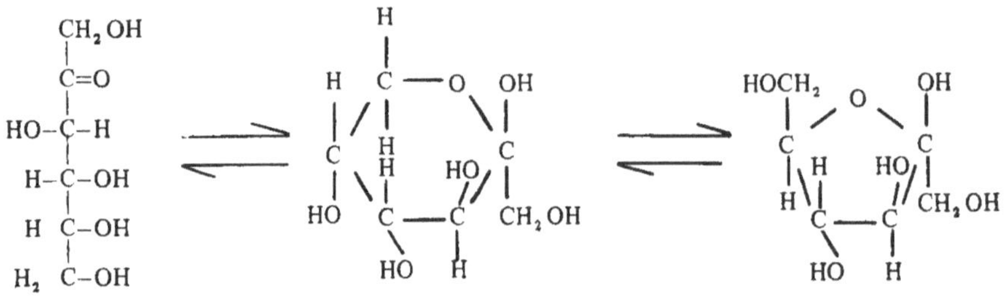


## II. PHYSICAL AND CHEMICAL PROPERTIES OF FRUCTOSE

Fructose is a highly reactive keto-hexose with the typical monosaccharide $C_6H_{12}O_6$ empirical formula. It crystallizes as anhydrous prisms of β-D-fructopyranose and has a melting point of 102 to 104°C. In solution, an equilibrium is established among the anomers. The much sweeter pyranose form dominates (67 to 31%) the furanose structure (Table 1). The open-chain keto configuration is present to a very small degree. However, in some reactions, this form manifests its presence, particularly during the heating of fructose solutions.[6]

Fructose is the most water-soluble of all sugars, a property advantageously utilized in dry mixes that must be rapidly dissolved in limited volumes of cold water prior to consumption. Table 2 shows a comparison of fructose solubility and that of sucrose and dextrose.

A typical assay for food-grade crystalline fructose that conforms to the Codex Alimentarius Draft International Standard (Alinorm 76/27) for fructose is

| | |
|---|---|
| Assay | Minimum 99.5% |
| Water (Karl Fischer method) | Maximum 0.1% |
| Ash | Maximum 0.01% |
| Glucose | Maximum 0.1% |
| Color (ICUMSA) | Maximum 30 |
| pH (10% solution) | 4.5–7.0 |
| Arsenic, lead, sulphur dioxide | not detectable |

## III. SWEETNESS OF FRUCTOSE

For processors of reduced calorie foods, undoubtedly the most interesting attribute of fructose is that it is the sweetest of the natural sugars (Table 3). Recent investigations show that under optimum conditions the relative sweetness of crystalline fructose is 170 compared to 100 for sucrose.[22,32] Thus, many full calorie products can be converted to their reduced calories-from-sugar counterparts by replacing each 100 lb of sucrose with approximately 60 lb of fructose. Similarly, 100 lb of glucose or dextrose could be replaced on a sweetness basis with about 50 lb of fructose.

It should be emphasized that the perceived fructose sweetness is a function of its concentration and the pH and temperature of the product at

TABLE 2

**Concentration of Various Sugars in Saturated Aqueous Solutions at Various Temperatures**

| Temperature (°C) | Fructose (%) | Dextrose (%) | Sucrose (%) |
|---|---|---|---|
| 20 | 78.94 | 47.11 | 66.60 |
| 30 | 81.54 | 54.64 | 68.18 |
| 40 | 84.34 | 61.84 | 70.01 |
| 50 | 86.94 | 70.91 | 72.04 |

From Vanninen, E., unpublished data, Suomen Sokeri, Oy, Helsinki, 1970.

TABLE 3

**Relative Sweetness of Various Substances**

| Material | Relative sweetness | Method |
|---|---|---|
| Saccharin, as sodium salt | 675 | 2% sucrose |
| Fructose | 1.7 | Threshold |
| Sucrose | 1.0 | Standard |
| Glucose | 0.62–1.00 | Varied sucrose |
| Mannitol | 0.57 | Threshold |
| Sorbitol | 0.54 | Threshold |
| Lactose | 0.31–0.37 | Varied sucrose |

From Dermer, O., The science of taste, *Proc. Okla. Acad. Sci.*, 27, 9, 1946. With permission.

the moment of consumption. Therefore, it is critical to evaluate fructose sweetness in combination with all the ingredients of a new formulation under specific consumption conditions to determine precisely what kind of replacement ratios can be established. The maximum relative fructose sweetness occurs in cold, dilute acid solutions.[15] A typical reduced calorie product possibly developed along these lines would be a fructose-sweetened lemon beverage base in which all sucrose, dextrose, or corn syrup solids present in the regular product would be replaced by 60% or less of crystalline fructose. Quantities of all other flavoring ingredients would be the same. Since this reformulated beverage base contains 40% less sweetener solids, reduced usage is necessary. Rapid fructose solubility and the fact that a lesser weight is involved reduce the time required to dissolve and mix the new beverage base.

Temperature is an important factor in fructose sweetness development. At higher temperatures, the equilibrium state of the various anomers favors the furanose configuration, less sweet than the pyranose form, probably as a result of the pyranose anomer being more suitable to the tongue's sweetness receptors. The equilibrium states of the anomers are a function of the temperature at the time of consumption.[16] Therefore, a fructose-sweetened cake or cookie, intended for cold consumption, will have a significantly higher sweetness at the time it is eaten than immediately after removal from the oven.

Since fructose solutions are significantly less viscous (Table 4) than those made with other sweeteners (particularly corn syrups), and changes in viscosity affect sweetness perception, it is advisable to evaluate the relative sweetness of syrups made with 50% less fructose in water only after the intended viscosity of the finished reduced calorie syrups is achieved through the addition of a thickener system.[3]

The ability of fructose to exert a synergistic effect on other sweeteners is also an important consideration when evaluating fructose applicability in reduced calorie foods, particularly those where the sweetener is the major ingredient and cost is a paramount issue.[33] Various combinations of fructose and sucrose exhibit sweetening values greater than the sum of each sugar's expected sweetening value. This suggests the possibility of blending fructose and sucrose to maximize their cumulative sweetening ability, reduce sugar calories, and minimize total sweetener costs. No published data are available on the most effective sugar combinations, but food technologists will most likely welcome the challenge since their findings could cause significant reductions in calories and sweetening costs.

TABLE 4

**Viscosity of 70% Aqueous Sugar Solutions**

| Temperature (°C) | Fructose (poise) | Sucrose (poise) | Glucose (poise) |
|---|---|---|---|
| -10 | 40.17 | – | – |
| -5 | 21.13 | – | – |
| 0 | 11.62 | 36.28 | – |
| 10 | 4.05 | 12.06 | – |
| 20 | 1.76 | 4.82 | 10.36 |
| 40 | 0.44 | 1.14 | – |

From Finnish State Technical Research Institute Data, Helsinki, 1973.

## IV. SOURCES OF FRUCTOSE

Prior to the development of our present systematic nomenclature for carbohydrates, the isolation and identification of specific sugars often resulted in labeling sugars with common names related to their raw material source. Glucose was known as "grape" sugar; lactose was called "milk" sugar; fructose, (because of its presence in many fruits and berries) was known as "fruit" sugar (Table 5). The discovery of these sugars occurred during the 18th and 19th centuries. Presently, the synonymous terminologies are used frequently to refer to these sugars in advertisements and in ingredient declarations on food products.

In 1792, the European pharmacist, Löwitz, while attempting to isolate dextrose, separated a sugar that resisted attempts to be crystallized. He reported this phenomenon. The term "fruit" sugar was first used by Mitscherlich in 1843 as a result of his investigations. Pure fructose was finally isolated from cane sugar in 1847 by Dubrunfaut, a chemist in France.[62] In the same year, while working with the carbohydrate, inulin (a nutrient-reserve polysaccharide made up of fructo-furanose units commonly found in the roots of many plants), investigators noted the presence of a strongly levorotary sugar – most likely fructose.

Commercially, the first practical processes for production of pure crystalline fructose used such

TABLE 5

**Average Fructose Content of Honey, Fruits, and Berries**

| | Fructose content – edible portions (%)[a] | % Total carbohydrate (%)[b] | Total carbohydrates as fructose (%) |
|---|---|---|---|
| Honey | 40.50 | 82.3 | 38.91 |
| Grapes | 6.53 | 15.7 | 35.49 |
| Apples | 5.93 | 14.5 | 38.01 |
| Bananas | 5.85 | 22.2 | 39.00 |
| Pears | 5.60 | 15.3 | 33.33 |
| Cherries | 5.35 | 17.4 | 27.29 |
| Blueberries | 3.28 | 15.3 | 19.52 |
| Blackberries | 2.74 | 12.9 | 17.68 |
| Oranges | 2.56 | 12.2 | 18.28 |
| Grapefruit | 2.26 | 10.6 | 19.48 |
| Strawberries | 2.13 | 8.4 | 21.09 |

[a]From Dako, D. Y., Trautner, K., and Somogyi, J. C., Der Glukose, Fruktose und Sacchararosegehalt verschiedener fruchte, *Schweiz. Med. Wochenshr.*, 100, 1970; Eheart, J. F. and Mason, B. S., *J. Am. Diet. Assoc.*, 50, 1976; Hardinge, M. G., Swarner, J. B., and Crooks, H., *J. Am. Diet. Assoc.*, 46, 1965.
[b]From Watt, B. K. and Merrill, A. L., *Composition of Foods, Handbook No. 8*, Table 1, U.S. Department of Agriculture, Washington, D.C., 1964.

inulin-rich sources as the Jerusalem artichoke, dahlia plants, or chicory roots as raw materials. This was an expensive and complicated process, so fructose was not available for large scale commercial use. Present technology employs invert sugar syrup as the basic fructose source. However, it is speculated that various types of enzyme-treated corn syrups may become alternative raw materials for fructose production.[17]

The most recent European processes utilize chromatographic principles and ion-exchange technology to achieve inversion, demineralization, and separation steps necessary to isolate the fructose portion of the inverted sugar solution. Further fructose purification is achieved during the lengthy crystallization operation. The Finnish process is the only one in which crystallization is made from a water solution rather than from less economical ethyl alcohol solutions.[20]

Prior to 1975, no serious consideration was given to wide scale industrial use of pure fructose in the U.S. because of the limited supply and high price of obtainable quantities. However, news of the success of European products sweetened with fructose, primarily in the crystalline state, reached the U.S. in 1975. The pace of research and development work accelerated due to the market entry of commercial quantities of high purity, competitively priced fructose from Finland, Germany, and France.

## V. PHYSICAL HANDLING OF CRYSTALLINE FRUCTOSE

Pure fructose (99.5%) is available in free flowing crystalline form similar to granulated cane or beet sugar. In general, fructose crystallized from a water solution (Finnish variety) has slightly larger crystals than that crystallized from an alcohol solution (German and French processes). The crystallizing operation is itself an additional purification step; therefore, the crystal form of commercially available fructose will usually assay purer than the 70% water solution. However, in most food applications, this difference is insignificant.

Unfortunately, confusion results from the fact that fructose proves to be an excellent humectant, and the literature reports often refer to its hygroscopicity.[6,24] Some processors of regular and reduced-calorie dry mixes hesitate to include fructose in their formulations because they want to maintain the free flowing character of their products. To them, hygroscopic means the possibility of rock-hard, lumpy mixes.

Therefore, it is important to understand that under certain severe conditions of humidity and temperature, fructose exposed to the atmosphere will attract moisture. However, crystalline fructose stored between 15 and 25°C with a relative humidity below 60% will encounter no problems. Undoubtedly, one of the most important deter-

TABLE 6

**The Effect of Relative Humidity on the Moisture Absorption of Crystalline Fructose at 20°C**

| Relative humidity | Length of Exposure (days) 1 | 2 | 5 | 7 | 9 | 12 | 16 | 20 | 26 |
|---|---|---|---|---|---|---|---|---|---|
| 33 | 0.02 | 0.01 | 0.02 | 0.02 | 0.01 | 0.01 | 0.01 | 0.01 | 0.01 |
| 48 | – | 0.03 | 0.03 | 0.03 | 0.03 | 0.03 | – | – | – |
| 55 | – | 0.06 | 0.06 | 0.06 | 0.06 | 0.05 | – | – | – |
| 60 | 0.10 | 0.12 | 0.13 | 0.13 | 0.13 | 0.13 | 0.13 | 0.13 | 0.13 |
| 66 | 0.32 | 0.52 | 1.12 | 1.51 | 1.87 | 2.45 | 3.16 | 3.84 | 4.89 |
| 72 | 0.68 | 1.33 | 3.34 | 4.66 | 6.01 | 7.91 | 10.3 | 12.2 | 14.7 |
| 76 | 0.91 | 1.78 | 4.40 | 6.14 | 7.86 | 10.3 | 12.8 | 14.9 | 17.7 |
| 82 | 1.27 | 2.49 | 6.14 | 8.67 | 11.0 | 13.7 | 16.9 | 19.7 | 23.8 |

*Note:* Amount of moisture absorbed is indicated as percentage of dry weight.

From Research report, Finnish State Technical Research Center, No. A-6759/73, Helsinki, 1973.

minants of product quality and shelf life of fructose-based, dry products is control of the relative humidity between 30 and 60% during all mixing and packaging operations. Table 6 shows the effect of relative humidity alterations on percentage of moisture absorbed.

During transportation and storage of bags or drums containing fructose, it is inevitable that some stacking pressure will occur, causing the crystalline fructose to pack together. However, the lumps are easily broken up, frequently by unstacking and handling encountered in normal food plant operations. Unopened containers remain unaffected by the temporarily elevated temperatures encountered during ordinary boxcar or truck shipment, provided that this period is not excessively long.

In dry mixes, presence of nonsugar ingredients such as starch and cocoa may keep the fructose free flowing. Otherwise, the usual food grade anticaking agents should be blended with fructose, in quantities up to 2% of the fructose weight whenever possible thus ensuring flowability of the finished product.

## VI. HEALTH FOOD POTENTIAL OF FRUCTOSE

### A. In Dietetic/Diabetic Foods

Juvenile-onset diabetes, the most severe form of diabetes mellitus, is a result of the inability of the beta cells of the pancreas to secrete sufficient insulin in response to glucose stimulation and to catalyze crucially necessary steps in the metabolic pathway.[29] The milder form of diabetes, the so-called adult-onset variety, is associated with obesity in 60 to 80% of the patients.[7] Recent studies indicate that (due to obesity) the fat cells, muscle tissues, and liver of mild diabetics become resistant to the required hormonal insulin action, thus impeding entrance of energy providing glucose.[4,10] Consequently, blood sugar rises to excessive levels and spills over into the kidneys causing cardiovascular, renal, and other problems associated with the diabetic condition.[11] Because fructose does not require insulin for its transportation into peripheral tissue cells, nor is insulin necessary for its phosphorylation in the liver where it is preferentially converted to glycogen, the development of special fructose-sweetened dietary products offers exciting possibilities for diabetics.[1,8,23] Recent medical research in Finland disclosed that patients with mild or well-controlled diabetes can consume 1.0 to 1.5 g of fructose per kilogram of body weight per day (2.5 to 3.8 oz/day for a 154-lb person) without negative effect on the balance of glucose metabolism.[8] The total daily caloric intake must be maintained at a constant level by reducing other carbohydrates to allow for fructose consumption.

At this point, it is strongly advised that processors of foods intended for the diabetic market proceed with extreme caution when making label references to the efficacy and suitability of such products for diabetics. Although such claims are prominently labeled on diabetic foods sold throughout Europe, the U.S. Food and Drug Administration (FDA) has disfavored fructose labeling claims concerning its metabolic properties relative to stabilizing blood sugar levels and insulin demand. In 1974, a recall action was instituted against retail packages of fructose because of label statements objectionable to FDA, despite similarity to those found on European packages.

Therefore, it seems prudent to predicate any label statements primarily on the fact that the increased sweetness of fructose permits achievement of an acceptable sweetening level with significantly less fructose as compared to original full-sugar formulations. Experience has shown that as much as 40 to 50% of the original sugar calories in a food product can be eliminated by replacing all of the sucrose, glucose, and lactose in a formulation with 50 to 60% of their weight of crystalline fructose. On a calculated basis, depending on sweetness of the replaced sugars, a reduction in sweetness from 10 to 20% might be expected. However, in practice, this reduction is often judged insignificant by taste panelists. It is postulated that the original, fully sweetened products are oversweetened or that persons aware of the reduced-caloric nature of the tested product, subconsciously lower their hedonic expectations. A definite plus for products solely sweetened with fructose is that the bitter aftertaste long associated with reduced calorie products sweetened with saccharin is absolutely eliminated. The resulting clean sweetness of fructose provides a definite advantage in achievement of consumer acceptance for reduced-calorie products.

Interestingly, fructose possesses a marvelous ability to mask the aftertaste of saccharin. A 99 to 1 mixture of fructose and sodium saccharin proved to be a superior concentrated sweetening system in low calorie soft drinks, powdered beverage mixes, and hot chocolate powders.[13,14] However, in view of the proposed ban on saccharin use in the U.S., such mixtures will no longer be permitted in this country. Fructose has also been effectively combined with the cyclamates, although at different ratios than those used with saccharin because of the cyclamates' lesser sweetening capacities.

If the FDA's ban on saccharin use becomes effective, no single, nonnutritive sweetener will be available to persons on restricted diets. Changes in the Delaney clause relative to the absolute noncarcinogenicity imperative of food ingredients may be made due to the furor over the saccharin ban. However, the value of fructose to processors of reduced-calorie foods is independent of any regulatory decisions regarding nonnutritive sweeteners. This fact should become increasingly apparent as public attention is directed to the legitimate place of fructose, the sweetest natural sweetener, in the diets of health-conscious individuals.

**B. In Foods for Hypoglycemics**

In some respects, hypoglycemia is the reverse side of the diabetic coin. Diabetes is characterized by an excess of blood glucose; hypoglycemia is a metabolic disorder in which the blood glucose is abnormally low.[30] The most common form of this condition is reactive hypoglycemia resulting from the rapid glucose assimilation achieved by an excessive insulin secretion from the beta cells of the pancreas.[26] This excessively rapid utilization of blood glucose reduces the circulating blood sugar to such low levels that the brain's blood supply is severely reduced, and the symptoms of hypoglycemia such as fatigue, nervousness, ravenous hunger, and shaking appear.[18] These symptoms result from the stress response initiated by a blood sugar deficiency, since blood sugar is the brain's sole energy source. The brain, unlike other body tissues, is unable to catabolize proteins or fats for energy.

Recently, "fasting diets" received considerable attention from persons endeavoring to reduce weight quickly. However, such a regime is often accompanied by anxiety and depression due to failure of the diet to provide sufficient blood sugar for the brain's energy demands. Processors of reduced-calorie foods should be aware of such potential problems when formulating their specialty products. By sweetening these low calorie foods with fructose, a balance can be achieved among the physiological demands for blood sugar, the desire to reduce calories, and the need to give the product an acceptable taste.[21]

## VII. FRUCTOSE AND DENTAL HEALTH

Due to increased educational efforts designed to improve the dental health of the U.S. population, the relationship of sugar and products containing sugar to dental caries has become an important incentive for the purchase of reduced-sugar foods by health-conscious consumers. Although the caloric reduction of such foods is a concomitant benefit, the primary emphasis has been on improved dental health aspects. Therefore, it is appropriate to examine fructose in relation to the incidence of dental caries.

Cariogenic bacteria in the oral cavity synthesize adhesive polysaccharides (dextrans) by acting on carbohydrates, particularly sucrose, and cause dental plaque.[19] This plaque serves as a growth medium for decay causing acidogenic bacteria in close proximity to the teeth. Because the type of polysaccharides (levans) resulting from oral enzymatic activity upon fructose are much more soluble than the dextrans, the formation of plaque and dental caries is significantly reduced when fructose is substituted for sucrose in a daily diet. These findings were confirmed in the comprehensive 2-year Turku Dental Studies[30] conducted as a part of the research evaluating the effect of sucrose, fructose, and xylitol on dental health.

Mindful of the above, it seems valid to conclude that a reduced-calorie product, sweetened entirely with fructose, would possess a double attraction for consumers concerned with dental health. Not only would the sugar quantity be lower, but the less harmful fructose would be present.[12] The substitution of fructose for sucrose on the product label will not go unappreciated by those consumers perceptive enough to read ingredient statements in the first place.

## VIII. REDUCED-CALORIE PRODUCT POSSIBILITIES

Desserts and dessert mixes are some of the most exciting areas for new product development using the calorie-sparing sweetness of fructose. By utilizing increased sweetening and humectant properties of pure fructose and by replacing all the usual sugar and syrups with an equivalent sweetness (calorically reduced) quantity of crystalline fructose, a complete line of premium quality, special diet cakes and cookies can be made. Since reductions of up to 50% of total sugars are possible, it is necessary to employ noncaloric bulking agents such as various types of cellulose, to achieve the required volume and texture of the finished baked goods.[27]

Reduced-calorie, complete cake mixes combining the attributes of fructose with those of an unsaturated fat system, a low sodium content, or with other dietetic food features can be made. Such mixes appeal to that large population section who literally want to eat their cake and have their diets too.

Frozen desserts intended for the dietetic/diabetic market encountered severe regulatory restriction in 1976 when the FDA acted against frozen desserts sweetened with a combination of sorbitol and saccharin. At that time, the FDA reaffirmed agency policy that combinations of nutritive and nonnutritive sweeteners were not permitted by the interim Food Additive Order for saccharin.[28] However, a very effective solution to this dilemma is available by replacement of sorbitol with an equal weight of crystalline fructose. Because fructose and sorbitol contribute the same number of calories to food and since their metabolic pathways after absorption are identical, it follows that the far greater sweetness of fructose can be utilized to make reduced-calorie frozen desserts without using the prohibited sorbitol-saccharin combination.[9]

When the new federal standards of identity for frozen desserts are published, certainly changes will be made permitting much greater freedom and technological creativity in the selection of safe, suitable ingredients used in the manufacture of ice cream and other rigidly standardized frozen desserts. Particularly interesting to producers of reduced-calorie varieties is the new requirement stated in terms of minimum protein levels rather than minimum percentages of total milk solids. Since the major portion of nonfat dry milk consists of lactose, having very low sweetening power, it may be possible to make a reduced-calorie frozen ice cream or iced milk by using a combination of high protein sodium caseinate and a greatly reduced sugar percentage in the form of super-sweet crystalline fructose replacing lactose.

Instant pudding powdered mixes and flavored gelatin dessert powders can be elegantly converted to their reduced-calorie counterparts by making them with fructose in an amount that is 50% of the original carbohydrate sweetener content. In gelatin desserts, use of powdered gelatin of the highest bloom strength with fructose produces a total caloric reduction approaching 50% without significantly reducing gelatin dessert flavor or eating qualities. In such applications, characteristic fruit, berry, and citrus flavors are perceived more distinctly. The somewhat diminished sweetness is considered insignificant because of increased overall flavor.

Processors of infant's and children's foods should definitely consider the merits of severely restricting sugar content and sugar calories to deter attacks of highly vocal opponents of sugars.

Unquestionably, by use of fructose, an entire line of adequately sweet baby foods and cereals could be efficiently produced and marketed despite the fact that they contain only half of their current quantity of added sugar. Fructose reduction of dental carries should be particularly welcomed by parents concerned with their children's dental health.

One of the most successful reduced-calorie carbonated beverages presently on the market is a lemon-flavored cola soft drink which derives half its sweetness from sucrose and the balance from a small quantity of sodium saccharin (about 0.02%). Because of the acid character of this drink and its compatible flavoring system, it is particularly adaptable to total or partial sweetening by fructose.[15] If the saccharin ban becomes effective, then a very acceptable version of this light-in-calories drink could be made by using a quantity of fructose (equal to about 7% by weight) in place of all sucrose and saccharin.

Reduced-calorie salad dressing mixes, particularly those of the pourable variety, can be made by blending spices, an emulsifying agent like xanthan gum, malic acid, and enough crystalline fructose to provide about 5% fructose in the finished dressing. Frequently, when low calorie dressings are made, caloric reductions are due primarily to a diminished vegetable oil percentage. Little attention is devoted to their often excessively high levels of sugar or corn syrup. Further reductions in total calories may be accomplished by use of fructose in a dry salad dressing mix while simultaneously improving flavor and dispersion characteristics.

Dietetic chocolate is a particularly interesting food in which candy technologists may combine their expertise with the unique features of fructose to make candies[34] that are less damaging to a diet. Chocolate or carob coatings sweetened exclusively with fructose have such superior flavor that thinner coatings can be utilized in enrobing operations. A carob-fructose bar should appeal greatly to those individuals who abhor chocolate-sugar confections. Coated or uncoated energy bars can be made utilizing the smooth energy release capabilities and the superior sweetness of fructose in a blend with high protein components.

Undoubtedly the simplest way to use fructose to make a reduced calorie food is to package crystalline fructose in individual amounts furnishing the sweetening equivalent of a level teaspoon of granulated sugar. Since a level teaspoon of sugar weighs about 5.9 g, a sweetening equivalent of fructose should weigh about 3.6 g. Fructose can easily mix with other sugars or noncaloric bulking agents and sell as a tabletop sweetener. To maintain flowability, it is advisable to blend in 0.5% silicon dioxide or similar anticaking agents and to package the product in a moisture-resistant, reclosable container. Storing fructose in a dry container at moderate temperatures insures a shelf life of several years.

## IX. ECONOMICS OF FRUCTOSE USE

Barring a technological miracle in fructose manufacture or a rise in U.S. sugar prices to levels higher than ever, the cost of crystalline fructose will always exceed that of an equivalent sweetening quantity of sucrose or glucose. The need to employ raw material sources similar to those required in refining sucrose plus the expense of extremely large energy amounts needed to produce crystalline fructose, make it highly unlikely that a fructose selling price ratio of less than 2.5 times that of sucrose will be possible. By 1980, a large scale domestic production of crystalline and 70% solutions of fructose will begin in the U.S. The 2.5 price multiple will be a possibility.

Therefore, the decision to use pure fructose in reduced-calorie foods must be based on criteria other than price. When all of the organoleptic, metabolic, and nutritional attributes are considered, it is likely that many processors of special diet and reduced-calorie foods will conclude that benefits of sweetening with fructose far exceed any additional costs.

# REFERENCES


1. **Akerblom, H. K., Siltanen, I., and Kallo, A. K.,** Does dietary fructose affect the control of diabetes in children? Symp. Clin. and Metab. Aspects of Fructose, *Acta Med. Scand.*, Suppl. 542, 195, 1972.
2. **Aminoff, C.,** New carbohydrate sweeteners, in *Sugars in Nutrition*, Sipple, H. L. and McNutt, K. W., Eds., Academic Press, New York, 1974, chap. 10.
3. **Arabie, P. and Moskowitz, H. R.,** The effects of viscosity upon perceived sweetness, *Percept. Psycophys.*, 9, 410, 1971.
4. **Brothers, M. S.,** *Diabetes, the New Approach*, Grosset and Dunlap, New York, 1976.
5. **Cantor, S.,** The chemistry and technology of sugar, in *Sugars in Nutrition*, Sipple, H. L. and McNutt, K. W., Eds., Academic Press, New York, 1974.
6. **Doty, T. and Vanninen, E.,** *Pure Fructose as an Industrial Raw Material*, Finnish Sugar Co., Helsinki, 1975.

6a. **Dubrunfaut, M.,** *Am. Chim. Phys.*, 21, 169, 1847.

7. **Felig, P.,** Pathophysiology of diabetes, in *Diabetes Mellitus*, 4th ed., American Diabetes Association, New York, 1975, 3.
8. Fructose, Fruit Sugar, A Review of Nutritional, Medical and Metabolic Aspects, Finnish Sugar Co., Helsinki, 1973.
9. **Gabbay, K. H.,** The sorbitol pathway in the nervous system, in *Sugars in Nutrition*, Sipple, H. L. and McNutt, K. W., Eds., Academic Press, New York, 1974.
10. **Heinz, F.,** Metabolism of fructose in liver, Symp. on Clin. Metab. Aspects of Fructose, *Acta. Med. Scand.*, Suppl. 542, 27, 1972.
11. **Herman, R. H.,** Intestinal metabolism of fructose, Symp. on Clin. Metab. Aspects of Fructose, *Acta. Med. Scand.*, Suppl. 542, 19, 1972.
12. **Huttunen, J. K.,** Effects of sucrose, fructose and xylitol diets on glucose, lipid and urate metabolism, Turku Sugar Studies, XI, *Acta Odontol. Scand.*, Suppl. 70, 001, 1975.
13. **Hyvonen, L.,** Results of Measuring the Sweetness of Fructose-Saccharine Mixtures, University of Helsinki, 1973 (unpublished).
14. **Hyvonen, L.,** The Masking of Saccharine After-taste in Foodstuffs, University of Helsinki, 1974 (unpublished).
15. **Hyvonen, L.,** The Sweetness of Crystalline and Just Dissolved Fructose, University of Helsinki, 1974, (unpublished).
16. **Hyvonen, L. and Koivistoinen, P.,** The Tautomeric Basis for Varying Sweetness of Fructose, paper presented at annual meeting of the Institute of Food Technologists, Anaheim, Cal., 1976.
17. **Kooi, E. R. and Smith, R. J.,** Dextrose-levulose syrups from dextrose, *Food Technol.*, 26, 57, 1972.
18. **Lefebvre, P. and Luyckx, A.,** Spontaneous and insulin-induced hypoglycemia, in *Diabetes Mellitus*, 4th ed., American Diabetes Association, New York, 1975.
19. **Makinen, K. K.,** Sugars and the formation of dental plaque, in *Sugars in Nutrition*, Sipple, H. L. and McNutt, K. W., Eds., Academic Press, New York, 1974.
20. **Melaja, H. A.,** Procedure for the separation of fructose from the glucose of invert sugar, U.S. Patent 3,692,582, 1972.
21. **Mazovetsky, A. G.,** Use of Fructose for Treatment of Diabetes Mellitus, paper presented at the Symposium on Fructose and Diabetes Mellitus – Metabolic Aspects, Moscow, 1976.
22. **Moskowitz, H.,** *Models of Additivity for Sugar Sweetness, Sensation and Measurement*, D. Reidel Publishing, Dordrecht, Holland, 1973, 379.
23. **Nikkila, E. A.,** Influence of dietary fructose on serum triglycerides in hypertriglyceridemia and diabetes, in *Sugars in Nutrition*, Sipple, H. L. and McNutt, K. W., Eds., Academic Press, New York, 1974.
24. **Norrish, R. S.,** *A Study of the Rate of Gain or Loss of Moisture from Sugar and Sugar Confectionery Products During Storage*, Bartholomew Press, Dorking, United Kingdom, 1969, chap. 1.
25. **Page, L. and Friend, B.,** Level of use of sugars in the United States, in *Sugars and Nutrition*, Sipple, H. L. and McNutt, K. W., Eds., Academic Press, New York, 1974. chap. 7.
26. **Palm, J.,** *Diet Away Your Stress, Tension, and Anxiety*, Doubleday & Co. New York, 1976.
27. **Pike, R. L. and Brown, M. L.,** *Nutrition: An Integrated Approach*, Wiley & Sons, New York, 1967, chap. 2.
28. Regulatory Letters Hit Frozen Desserts Sweetened with Saccharin, *Food Chem. News*, p. 25, March 29, 1976.
29. **Roch-Norlund, A. E., Hultman, E., and Nilsson, L. H.,** Metabolism of fructose in diabetes, Symp. on Clin. Metab. Aspects of Fructose, *Acta. Med. Scand.*, Suppl. 542, 181, 1972.
30. **Scheinin, A., Mäkinen, K. K., and Ylitalo, K.,** Turku Sugar Studies: I, An intermediate report on the effect of sucrose, fructose and xylitol on the caries incidence in man, *Acta Odontol. Scand.*, 32, 383, 1974.
31. **Steinke, J.,** Hypoglycemia, in *Joslins Diabetes Mellitus*, 11th ed., Lea & Febiger, Philadelphia, 797, 1971.
32. **Shallenberger, R. S.,** Hydrogen bonding and the varying sweetness of the sugars, *J. Food Sci.*, 28, 584, 1963.
33. **Yamaguchi,** Studies on the taste of some sweet substances, *Agric. Biol. Chem.*, 34, 181, 1970.
34. **Zimmerman, M.,** Fructose in the Manufacture of Dietetic Chocolate, *The Manufacturing Confectioner*, p. 39, August 1974.

# INDEX

## A

## B

## C

## D

## E



## F

## G

## H



## I



## J



## K

# O



# P

## T

## U



## V



## W



## X



## Y